Advances in Intelligent Systems and Computing

Volume 599

Series editor

About this Series

The series "Advances in Intelligent Systems and Computing" contains publications on theory, applications, and design methods of Intelligent Systems and Intelligent Computing. Virtually all disciplines such as engineering, natural sciences, computer and information science, ICT, economics, business, e-commerce, environment, healthcare, life science are covered. The list of topics spans all the areas of modern intelligent systems and computing.

The publications within "Advances in Intelligent Systems and Computing" are primarily textbooks and proceedings of important conferences, symposia and congresses. They cover significant recent developments in the field, both of a foundational and applicable character. An important characteristic feature of the series is the short publication time and world-wide distribution. This permits a rapid and broad dissemination of research results.

More information about this series at http://www.springer.com/series/11156

Paul Fechtelkotter · Michael Legatt
Editors

Advances in Human Factors in Energy: Oil, Gas, Nuclear and Electric Power Industries

Proceedings of the AHFE 2017 International Conference on Human Factors in Energy: Oil, Gas, Nuclear and Electric Power Industries, July 17–21, 2017, The Westin Bonaventure Hotel, Los Angeles, California, USA

Springer

Editors
Paul Fechtelkotter
IBM Corporation
Medfield, MA
USA

Michael Legatt
Electric Reliability Council of Texas, Inc.
Taylor, TX
USA

ISSN 2194-5357 ISSN 2194-5365 (electronic)
Advances in Intelligent Systems and Computing
ISBN 978-3-319-60203-5 ISBN 978-3-319-60204-2 (eBook)
DOI 10.1007/978-3-319-60204-2

Library of Congress Control Number: 2017943046

Printed on acid-free paper

This Springer imprint is published by Springer Nature
The registered company is Springer International Publishing AG
The registered company address is: Gewerbestrasse 11, 6330 Cham, Switzerland

Advances in Human Factors and Ergonomics 2017

AHFE 2017 Series Editors

Tareq Z. Ahram, Florida, USA
Waldemar Karwowski, Florida, USA

8th International Conference on Applied Human Factors and Ergonomics and the Affiliated Conferences

Proceedings of the AHFE 2017 International Conference on Human Factors in Energy: Oil, Gas, Nuclear and Electric Power Industries, July 17–21, 2017, The Westin Bonaventure Hotel, Los Angeles, California, USA

Advances in Affective and Pleasurable Design	*WonJoon Chung and Cliff (Sungsoo) Shin*
Advances in Neuroergonomics and Cognitive Engineering	*Carryl Baldwin*
Advances in Design for Inclusion	*Giuseppe Di Bucchianico and Pete Kercher*
Advances in Ergonomics in Design	*Francisco Rebelo and Marcelo Soares*
Advances in Human Error, Reliability, Resilience, and Performance	*Ronald L. Boring*
Advances in Human Factors and Ergonomics in Healthcare and Medical Devices	*Vincent G. Duffy and Nancy Lightner*
Advances in Human Factors in Simulation and Modeling	*Daniel N. Cassenti*
Advances in Human Factors and System Interactions	*Isabel L. Nunes*
Advances in Human Factors in Cybersecurity	*Denise Nicholson*
Advances in Human Factors, Business Management and Leadership	*Jussi Kantola, Tibor Barath and Salman Nazir*
Advances in Human Factors in Robots and Unmanned Systems	*Jessie Chen*
Advances in Human Factors in Training, Education, and Learning Sciences	*Terence Andre*
Advances in Human Aspects of Transportation	*Neville A. Stanton*

(continued)

(continued)

Advances in Human Factors, Software, and Systems Engineering	*Tareq Z. Ahram and Waldemar Karwowski*
Advances in Human Factors in Energy: Oil, Gas, Nuclear and Electric Power Industries	*Paul Fechtelkotter and Michael Legatt*
Advances in Human Factors, Sustainable Urban Planning and Infrastructure	*Jerzy Charytonowicz*
Advances in the Human Side of Service Engineering	*Louis E. Freund and Wojciech Cellary*
Advances in Physical Ergonomics and Human Factors	*Ravindra Goonetilleke and Waldemar Karwowski*
Advances in Human Factors in Sports, Injury Prevention and Outdoor Recreation	*Tareq Z. Ahram*
Advances in Safety Management and Human Factors	*Pedro Arezes*
Advances in Social & Occupational Ergonomics	*Richard Goossens*
Advances in Ergonomics of Manufacturing: Managing the Enterprise of the Future	*Stefan Trzcielinski*
Advances in Usability and User Experience	*Tareq Ahram and Christianne Falcão*
Advances in Human Factors in Wearable Technologies and Game Design	*Tareq Ahram and Christianne Falcão*
Advances in Communication of Design	*Amic G. Ho*
Advances in Cross-Cultural Decision Making	*Mark Hoffman*

Preface

Human factors in energy focus on the oil, gas, nuclear, and electric power industries that aim to address the critical application of human factors' knowledge to the design, construction, and operation of oil and gas assets, to ensure that systems are designed in a way that optimizes human performance and minimizes risks to health, personal or process safety, or environmental performance. This book focuses on delivering significant value to the design and operation of both onshore and offshore facilities. Energy companies study the role of human behavior for safety and accident prevention; however, third party providers and different operators have different standards and different expectations. While oil and gas exploration and production activities are carried out in hazardous environments in many parts of the world, offshore engineers are increasingly taking human factors into account when designing oil and gas equipment. Human factors in fields such as machinery design, facility and accommodation layout, and the organization of work activities have been systematically considered over the past twenty years on a limited number of offshore facility design projects to minimize the occupational risks to personnel, support operations and maintenance tasks, and improve personnel well-being. Better understanding for human factors' issues also supports the nuclear industry's move from analog to digital control rooms. Human considerations, such as lighting, temperature, and even ergonomics, play important parts in the design. This book will be of special value to a large variety of professionals, researchers, and students in the broad field of energy modeling and human performance.

This book will be of special value to a large variety of professionals, researchers, and students in the broad field of energy research and human performance who are interested in situation awareness, training, and intelligent operator support. We hope this book is informative, but even more that it is thought provoking. We hope it inspires, leading the reader to contemplate other questions, applications, and potential energy solutions in creating good designs for all.

Our sincere thanks and appreciation go to the Board members listed below for their contribution to the high scientific standard maintained in developing this book.

July 2017

Paul Fechtelkotter
Michael Legatt

Contents

When is a Human Factors Review Appropriate? Development of a Human Factors Screening Tool for Nuclear Regulatory Commission Project Managers

Brian D. Green(✉)

U.S. Nuclear Regulatory Commission,
11555 Rockville Pike, Rockville, MD 20852, USA
brian.green@nrc.gov

Abstract. Human factors experts in many industries experience challenges communicating the importance of human performance issues with people from outside the human factors community. Human factors staff at the Nuclear Regulatory Commission (NRC) are addressing this situation by conducting outreach activities to ensure that NRC project managers understand the scope of human factors activities. Human factors principles were used to create a decision-support tool to help NRC project managers promptly and accurately identify human factors issues in license applications. Since the rollout and implementation of the tool, there has been a noticeable increase in prompt and accurate routing of licensing actions to the human factors experts. A description of the process used to create the desk guide and a summary of outreach activities is included in the hopes that other organizations may achieve similar results.

Keywords: Human factors · Project management · Screening tool · Desk guide · Job aid · Outreach

1 Introduction

The Nuclear Regulatory Commission (NRC) regulates the civilian use of nuclear power. This is done through a variety of different processes and affects many different industries. Perhaps most notable, is the licensing of nuclear reactors. Organizations that wish to construct a new nuclear power plant, or modify an existing nuclear plant, must formally submit various licensing documents for NRC approval. These license applications are then reviewed by subject matter experts in a variety of technical domains to ensure that the design of the plant will not jeopardize the safety or security of the public. One of the technical areas of review is human factors. In order to ensure that operators can safely operate the plant, a human factors review assesses the characteristics of the design that may affect human performance such as the human-system interfaces, procedures, training, organizational structures and others.

Like many organizations, the NRC uses project managers to assign and manage large-scale work projects. One function of project managers at the NRC is to identify appropriate cognizant engineers to review specific issues found in license amendment

P. Fechtelkotter and M. Legatt (eds.), *Advances in Human Factors in Energy: Oil, Gas, Nuclear and Electric Power Industries*, Advances in Intelligent Systems and Computing 599, DOI 10.1007/978-3-319-60204-2_1

applications. This includes identifying the subset of licensing actions that need to be reviewed by human factors engineers and subsequently routing these license application materials to the appropriate organizations. Not all licensing actions require human factors review; and correctly identifying those that do require it is often not entirely straightforward.

Project managers screen all incoming licensing actions to identify issues that need to be reviewed. When a project manager identifies a human factors issue he/she will contact the human factors group to confirm that they should be included in the review. If the human factors team agrees, then the formal human factors review process is started. The human factors team may also decide that a review is not necessary, in which case no human factors review will be conducted.

This process depends, in part, upon the project manager correctly identifying human factors issues. The degree of familiarity with human factors concepts varies greatly within the project management organization; some project managers have significant human factors experience, while others have little or none. The lack of understanding of human factors issues coupled with the inherently complex nature of human factors principles can make proper identification of license applications a difficult and error prone task. There are several tools available to help project managers decide which groups should review a particular license application. These tools work well most of the time, however, there are occasional failures.

The timing of identification of issues is important. For the sake of this discussion, "prompt identification," is defined as when the project manager correctly identifies the need for a human factors review when one is necessary during his or her initial screening of a license application. In contrast, "late identification" is any time the need for a human factors review is identified *after* the initial screening. This can occur as a result of a variety of processes intended to ensure that license applications do not miss review from cognizant engineers.

It is unclear if there are instances when a human factors review was missed that lead to safety consequences at a nuclear plant.[1] However, there have been multiple instances where late identified issues have caused inefficiencies in the review process, adding additional cost and frustration to both the NRC and licensees/applicants. Therefore, staff in the Probabilistic Risk Assessment Operations and Human Factors Branch (APHB) of the Office of Nuclear Reactor Regulations (NRR) applied human factors concepts (signal detection theory, and basic job-aid development principles) to develop a strategy to minimize the likelihood of late identified and/or missed opportunities for human factors review of appropriate license applications.

Failing to promptly and correctly identify the appropriate cognizant reviewers can have negative consequences including: (1) slippage of NRC and licensee schedules, (2) displacement of other work when reviewers have to adjust their schedules to accommodate unexpected reviews, (3) establishment of poor precedent if licenses are amended without proper assessment (i.e. reviewers may face certain legal challenges

[1] The NRC did not conduct human factors reviews prior to the Three Mile Island accident of 1979. So, although this was an event caused, in part, by human factors considerations, this would not be a relevant example to the current discussion because the review process at the time was not comparable to the process used today.

when correcting precedent during subsequent reviews), (4) potential negative consequences if issues are identified by other reviewers or advisory groups during late stages of the reviews (such as the Advisory Committee for Reactor Safeguards or external stakeholders), and (5) potential safety issues at plants if license amendments are approved and implemented without appropriate review.

A desk guide based on existing NRC human factors guidance was developed to help NRC project managers quickly and accurately identify those licensing actions that may need to be reviewed by human factors experts. A description of the human factors principles used to develop and implement the desk guide are included below. The desk guide is currently being used to improve communications between the human factors staff and the project management organization.

Challenges communicating human factors issues are not unique to the NRC. The methods described in this paper demonstrate how human factors principles can be used to conduct outreach activities with people who are unfamiliar with human factors principles. This method has the potential to be adapted to other organizations and applications. The use of screening questions, like those found on our desk guide, may facilitate useful discussions about how to effectively communicate about the breadth of scope of human factors work.

2 Method

2.1 Goal

The goal of this project was to create an intervention that would help ensure that project managers promptly and accurately identify human factors issues in licensing processes, thereby reducing any delays and costs caused by late routing and preventing approval of licensing actions without review of human factors staff. Any intervention developed had to be low cost and require minimal effort to implement. Also, the intervention needed to provide obvious value to the project management organization and be able to work across organizational boundaries because the human factors group has no authority over how the project management organization conducts business.

2.2 Theoretical Considerations

It was suspected from the beginning of this project that the problem described above was a signal detection issue. Some project managers were relatively unaware of what human factors issues looked like, and were therefore unable to identify them when they were present in licensing actions. Other project managers were familiar with routine human factors issues, but had difficulty identifying novel issues.

This problem can be defined in terms of signal detection theory [1]. In this case, a hit is when the project manager correctly identifies that a human factors review is necessary when there are human factors issues in a license application *during the initial screening process*. A miss is defined as when the project manager fails to identify a human factors issue *during the screening process* when there is a human factors issue in the application (i.e. late identification or no identification). A false alarm occurs when the project manager thinks there is a human factors issue, but after a brief review, human factors staff determines that the issue does not need NRC review (such as when there is no regulatory basis to review the issue). Finally, a correct rejection occurs when the project manager correctly decides that there are no human factors issues that need to be reviewed. Table 1 below illustrates the signal detection conditions.

Table 1. Identifying human factors issues in license reviews is a signal detection task. This table defines the signal detection conditions for this task.

		State of the world: Human Factors (HF) issues in license amendment?	
		HF review needed	HF review NOT needed
Project Manager (PM) behavior	PM promptly requests HF review	Hit	False alarm
	PM does NOT promptly request HF review	Miss	Correct rejection

Careful consideration was given to the costs associated with errors (misses and false alarms). The costs of a single miss could be severe, including schedule delays and potential safety implications. On the other hand, the cost of a single false alarm was minimal: a human factors reviewer would review the application, come to the conclusion no review was necessary, and document his/her decision. This process typically takes only 15–30 min. However, the cost of many false alarms has the potential to accumulate, thus robbing time away from the amount of time available for actual technical reviews. This meant that an increase in project manager sensitivity (*d'*)[2] was preferable to an increase in project manager response bias (beta).[3] An increase in beta would be expected to increase both the number of hits and false alarms, while an increase in *d'* would lead to an increase of hits without a significant concurrent increase in false alarms. Furthermore, a moderate increase in false alarms was deemed acceptable as long as APHB was not overwhelmed with false alarms.

2.3 Development of Desk Guide

Training can be an effective means of increasing sensitivity in some signal detection tasks [2]. Therefore, it was hoped that a desk guide, paired with a brief training session, could be an effective way of improving project manager sensitivity.

[2] Increasing sensitivity (or *d'*) would be indicative of increasing the ability of the project manager to distinguish human factor issues from non-human factors issues (noise). We would expect the number of hits to increase (and misses to decrease) with a minimal effect on the number of false alarms.

[3] Increasing the response bias (or beta) would increase the probably of the project manager saying "yes, a human factors review is necessary". We would expect both the number of hits and false alarms to increase. Misses would also decrease, but an additional cost is incurred as false alarms increase.

Figure 1 shows the front of the desk guide. The two-sided, half-sheet card has a diagram illustrating common human factors processes on the front that span the breadth of NRC human factors activities as well as a brief written summary of human factors activities. This helps to show readers the wide breadth of activities that human factors reviews affect. In addition, there is also a brief description of how the card should be used.

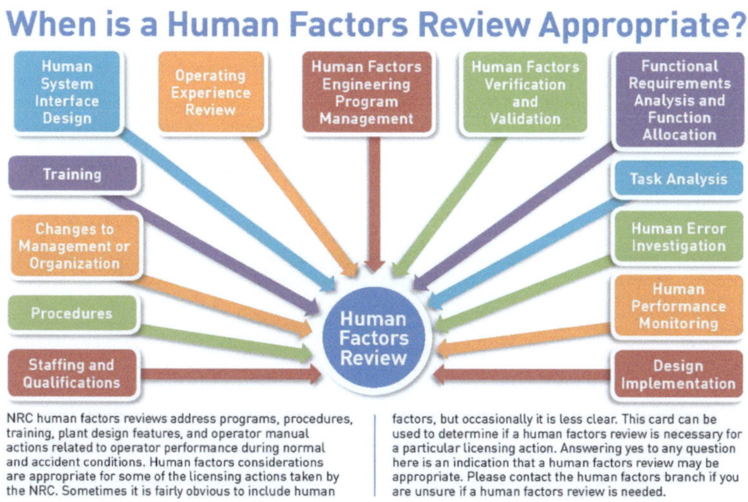

Fig. 1. The front of the desk guide provides a high-level overview of several human factors related processes [3].

The back of the desk guide (see Fig. 2) has several "yes/no" questions that project managers can use to screen licensing actions for human factors reviews. Answering "yes" to any question indicates that human factors experts should be consulted. Human factors technical reviewers then decide whether the application needs a formal review or not.

The screening questions were developed by combining, simplifying, and reducing information in several NRC human factors related guidance documents [4–7]. These guidance documents consist of hundreds of pages of human factors concepts and review criteria used in assessing a variety of licensing applications. They are available to all project managers, however, it is not expected that the project managers will have read these documents in their entirety or that they will have in-depth knowledge of the material in them. Human factors experts use this guidance regularly and have a technical background consistent with the technical material in these documents. Therefore, it was necessary to reduce the scope of the screening questions and strip away human factors jargon to ensure that all project managers would be able to use the desk guide.

To do this, a systematic approach was employed. An initial set of screening questions was reviewed by senior human factors staff who then added questions related

Does the License Amendment or Exemption Request:

- ✓ Propose changes to the controls, displays, alarms, or tools in the control room, local workstations, remote shutdown workstation, or the technical support center?
- ✓ Add, delete, or modify operator manual actions?
- ✓ Include time-critical manual actions?
- ✓ Cause changes in the operator skills, knowledge, or abilities needed to successfully complete manual actions?
- ✓ Change the amount of time needed or available to complete a manual action?
- ✓ Have the potential for performance shaping factors (such as stress, lighting, communication, task complexity, ergonomics, etc.) to influence the reliability of manual actions?
- ✓ Propose changes to the control room, human-system interfaces, procedures, or training that increase operator workload?
- ✓ Use automation to replace a formerly manual action?
- ✓ Use a manual action to replace a formerly automated action?
- ✓ Include secondary tasks (such as emergency declaration notifications or fire brigade duties) that are likely to divert operator attention or other resources away from important manual actions?
- ✓ Cause changes to the number of staff available to perform tasks?
- ✓ Resemble licensing actions from other facilities that have experienced human performance problems as a result of similar changes?
- ✓ Propose using components, tools, displays, etc. for a purpose other than the purpose for which it was designed?
- ✓ Include permanent or temporary compensatory actions to address issues with the facility?
- ✓ Influence operator fatigue in a negative way. Note that fatigue occurs on multiple levels including: 1) acute fatigue caused by several repetitions of an action within a short period of time, 2) long work shifts, or 3) sustained stressful activities over several weeks or months.
- ✓ Involve leaving non-functioning legacy equipment installed next to functioning new equipment or having old and new systems both functioning simultaneously?
- ✓ Have the potential for increased human error due to modifications planned while the unit is online/offline?
- ✓ Add, delete, or change procedures or technical specifications?
- ✓ Include human actions that are likely to change over time (such as due to future planned modifications or infrequently performed operations that may be forgotten)?
- ✓ Change the management organization in a way that may prevent human performance issues from being completely addressed or in a way that may affect safety culture?
- ✓ Need to be submitted to resolve a human performance Licensee Event Report (LER)?
- ✓ Relate to a 10 CFR Part 21 defect report on a human-system interface or vendor manual/instructions?

Fig. 2. The back of the desk guide has several yes/no screening questions that can be used to help project managers decide if human factors experts should review a particular licensing action [3].

to other, less frequently used processes. A draft version of the desk guide was then reviewed by a branch chief, NRC project manager, and division management to ensure that the language used was sufficiently free of human factors jargon that may be confusing to non-human factors staff.

2.4 Implementation and Outreach

Several methods were used to bring awareness of the problem of late identified reviews and to propose solutions.

APHB started sharing draft versions of the desk guide with various levels of management whenever a late identified review was identified in practice. A description of the problem was provided and APHB offered to give presentations about the scope of human factors reviews.

In September of 2015, APHB provided a formal presentation to the NRR project management organization. This presentation: (1) gave a high-level introduction to what the field human factors really is and addressed some common misperceptions, (2) provided examples of licensing actions that had human factors issues that were not obvious to most project managers, (3) described the costs associated with late identification of human factors issues in the licensing process, (4) introduced the desk guide, and (5) had a question and answer session so that project managers could learn from their own experiences.

In addition, a digital version of the desk guide was made available to all NRC staff via a SharePoint page and in the Agencywide Documents Access and Management System (ADAMS), the NRC document management system which is available to the public. This increased the probability that staff seeking more information about human factors would find the guidance. For instance, a project manager seeking more information about human factors might come across the desk guide as a result of a search query.

The desk guide has also been shared informally with organizations internal to the NRC in the hopes of communicating human factors issues project managers around the agency.

3 Results

Unfortunately, formal metrics do not exist to estimate the number of misses that occur during reviews so it is impossible to quantify the impact of the desk guide. However, the number of false alarms has increased modestly, so it is likely that the number of hits has also increased. Although it is impossible to conclude with certainty that the number of misses has been reduced, it is highly likely because of the apparent increase in beta.

In addition to the apparent change in hit/false alarm rate, there was an increased number of conversations about human factors, the breadth and scope of reviews, and technical details (especially for first of a kind reviews). These conversations were typically informal, but provided excellent opportunities to continue educating project managers about the value of human factors.

4 Discussion

Since the formal roll out of the desk guide in September of 2015, there has been a noticeable increase in engagement between the project managers and APHB human factors staff. As expected, a dramatic increase in screening requests was observed. The increase in false alarms was likely indicative of a reduction in misses, and was not so dramatic as to significantly distract staff from technical review duties. Therefore, the intervention was considered a success.

Several positive qualitative differences were observed. First, an increased dialog was observed between project managers and human factors experts. Detailed technical conversations began to occur as part of the screening process, rather than during the formal technical review process. This is contrary to what happened prior to the intervention (project managers often read the application independently and routed the review based upon their own experience.) These informal technical conversations allow the human factors experts to help decide if a human factors technical review is necessary, thus increasing the accuracy of the process.

These informal conversations may also have secondary benefits. They provide opportunities to educate the project managers about the human factors issues. Project managers now have the opportunity to learn about human factors issues as they are exposed to the less obvious human factors issues in reviews. These educational opportunities have the potential to further improve the sensitivity (*d'*) of project managers. These discussions also help the human factors staff remain cognizant of important licensing actions going on around the agency. Not all licensing reviews require human factors engagement, but the human factors staff can remain in the loop about agency activities outside the human factors domain.

There may be an additional benefit. The NRC has a focus on staff fungibility, in being able to move project managers from one organization to another. It is advantageous, from an organizational perspective to be able to take project managers from one office and reassigning them to another office with a different mission. Teaching NRR project managers about human factors can have positive effects on other NRC offices when the project managers are reassigned and take this new knowledge with them. This intervention may eventually increase the level of human factors knowledge and awareness throughout the agency.

However, nothing comes without cost. As expected, there was a noticeable increase of the number of screening requests since the intervention. Human factors staff must spend time getting up to speed with a particular licensing action to determine if an amendment requires human factors review. Often, this can be done in just a few moments (such as for routine license applications). In this case, the cost is minimal; the reviewer simply reads the introductory material, and communicates the need for a formal review to the project manager. However, when novel license amendments are received, this process can become much more labor intensive. In these cases, reviewers can begin by completing word searches for common human factors terms (such as: controls, alarms, displays, manual action, etc.) If the words are found, the reviewer can read the surrounding text to make an initial screening determination. Sometimes however, these types of words cannot be found and the reviewer must read large portions of application to make a determination. Applications vary in length significantly from less than a few dozen pages to several hundred. Luckily, the increase in the number of false alarms, has not been so extensive that they have detracted significantly from review work; therefore, they can be tolerated.

5 Conclusions

The NRC is guided by the following mission statement: "The NRC licenses and regulates the Nation's civilian use of radioactive materials to protect public health and safety, promote the common defense and security, and protect the environment" [8]. Accordingly, NRC technical reviewers (including human factors technical reviewers) focus on design issues that affect safety/security of nuclear plants (and other uses of radiation). This mission has served the NRC well for many years, and while opinions may vary about the appropriateness of the scope, we believe it is reasonable and appropriate for a regulator to focus on safety and security.

However, the broader discipline of human factors is capable of addressing aspects of workplace design that go well beyond the scope of the NRC mission (for instance, designing for efficiency, user enjoyment, reduction of workplace injuries, etc.) As a result of this limitation of scope, NRC human factors staff rarely get to demonstrate skills and methods that are outside of the scope. Therefore, it is possible that some NRC staff are completely unaware that human factors staff possess these types of skills.

Improving the efficiency of NRC internal work processes, ultimately helps staff serve the goal of protecting the safety and security of the public. Therefore, it is important to occasionally turn the metaphorical magnifying glass toward ourselves to ensure that our work processes are well calibrated. Human factors insights and perspectives can be instrumental to this process. This paper illustrates one way of doing so.

On September 15, 2015, the White House issued an Executive Order instructing government workers to use behavioral science insights to better serve the American people [9]. Although the order does not specifically mention human factors, it is clear that human factors principles and methods are consistent with the intention of the order. This paper illustrates one way in which human factors staff at the NRC have applied human factors principles to improve their own work processes to ultimately benefit the American people. Although the activities described in this paper were already under way when this Order was issued, we feel that this work demonstrates one example of how behavioral science insights can benefit the American people.

It's not usual for human factors experts to feel like their work is misunderstood or under-valued. By applying our human factors skills to our own work processes we have the potential to improve our own working conditions, and perhaps more importantly, improve the experience with our consumers, clients, licensees, members of the public, and others with whom we interact. We hope that other organizations will benefit from similar outreach activities that turn the human factors practices and principles back upon their own organizations. This gives co-workers a chance to see these skills in a new light, and an opportunity to witness first-hand the type of process improvements that are possible. In this sense, we can all use human factors to improve our own work processes and continue to build the credibility of the field.

Acknowledgments. The positions described in this paper are those of the author and do not necessarily represent the views of the U.S. Nuclear Regulatory Commission or the U.S. Government. Special thanks to Joe Giitter, Sam Lee, Sunil Weerakkody, Aida Rivera-Varona, George Lapinsky, Steven Lynch, and Niav Hughes for your support and for reviewing and improving the desk guide.

References

1. Green, D.M., Swets, J.A.: Signal Detection Theory and Psychophysics. Wiley, New York (1966)
2. Macmillan, N.A., Creelman, C.D.: Detection Theory – A User's Guide, 2nd edn. Lawerence Erlbaum Associates, New York (2005)
3. Green, B.D.: When is a Human Factors Review Appropriate? A Desk-Guide. U.S. Nuclear Regulatory Commission, Washington, DC. ADAMS Accession No. ML16020A423 (2016)
4. Nuclear Regulatory Commission: Standard Review Plan, Chapter 18: Human Factors Engineering (NUREG-0800, Chapter 18, Revision 2). U.S. Nuclear Regulatory Commission, Washington, DC (2007)

5. O'Hara, J.M., Higgins, J.C., Fleger, S.A., Pieringer, P.A.: Human Factors Engineering Program Review Model (NUREG-0711 Revision 3). U.S. Nuclear Regulatory Commission (2012)
6. O'Hara, J.M., Brown, W.S., Lewis, P.M., Persensky, J.J.: Human-System Interface Design Review Guidelines (NUREG-0700 Revision 2). U.S. Nuclear Regulatory Commission (2002)
7. Higgins, J.C., O'Hara, J.M., Lewis, P.M., Persensky, J.J., Bongarra, J.P., Cooper, S.E., Parry, G.W.: Guidance for the Review of Changes to Human Actions (NUREG-1764 Revision 1). U.S. Nuclear Regulatory Commission (2007)
8. Nuclear Regulatory Commission. https://www.nrc.gov/about-nrc.html
9. Executive Order No. 13707: Using Behavioral Science Insights to Better Serve the American People. 3 C.F.R. 56365-56367 (2015)

A GPP-Based Sectionalization Toward a Fast Power Transmission System Restoration

Saeedeh Abbasi[1], Masoud Barati[2], and Gino Lim[1(✉)]

[1] Industrial Engineering Department, University of Houston, Houston, TX 77204, USA
{sabbasi5, ginolim}@uh.edu

[2] Electrical and Computer Engineering Department, University of Houston, Houston, TX 77004, USA
mbarati@uh.edu

Abstract. High voltage transmission lines, in outdoor area, are in danger of extreme events such as tornadoes and hurricanes. Accordingly, terrible damage of transmission lines will cause a power grid blackout. Sectionalization as a part of a restoration process can make a power grid resilient by splitting it into multiple smaller areas. Then a diminutive portion of the total load is supplied at each area by black-start (BS) generation units with their self-start capability. To find the optimal sectionalization and perform a fast consumer electrification, a mathematical model is designed upon the association between the power transmission system sectionalization (PTSS) and graph-partitioning problem (GPP). The proposed GPP-based PTSS model finds the optimal sectionalization and restoration plan through a bi-level programming structure with sectionalization and restoration levels. Furthermore, pre-emptive goal programming (PEGP) supports the multiple objective termsof both levels. The model's efficiency is analyzed by IEEE 14- and 118-bus test systems.

Keywords: Graph partitioning problem · Power system restoration · Resilience · Sectionalization

1 Nomenclature

Sets		Parameters	
g	Index for generators, $g = \{1, \ldots, NG\}$	$\boldsymbol{\omega}$	Restoration time cost rate vector
m	Index for sections, $m = \{1, \ldots, NG\}$	$\mathbf{c}$	Marginal cost of generators
$b, \acute{b}$	Index for buses, $b = \{1, \ldots, NB\}$	**KG**	Bus-unit incident matrix
d	Index for demand loads, $d = \{1, \ldots, ND\}$	**KL**	Bus-line incident matrix
t	Index for time, $t = \{1, \ldots, NT\}$	**KD**	Bus-demand incident matrix
l	Index for transmission lines, $l = \{1, \ldots, NL\}$	**D**	Real power demand matrix

P. Fechtelkotter and M. Legatt (eds.), *Advances in Human Factors in Energy: Oil, Gas, Nuclear and Electric Power Industries*, Advances in Intelligent Systems and Computing 599, DOI 10.1007/978-3-319-60204-2_2

Parameters		Variables	
$P_g^{G,min}/P_g^{G,max}$	Maximal and minimal generating capacity of generation unit g	P_{gt}^{G}	Generated power of unit g at time t
RU_g^G/RD_g^G	Ramp up/down rate of unit g	$\mathbf{LS}/\mathbf{L\acute{S}}$	Load shedding matrix per demand/section (hourly)
$P_l^{L,max}$	Power line capacity of line l	s_{bm}	State of bus b at section m
$\mathbf{T}$	Total restoration time matrix	P_{lt}^{L}	Power flow on line l at time t
x_l	Reactance of line l	θ_{bt}	Phase angle of bus b at time t
$VOLL$	Value of loss of load	T_d^{Load}	Load pick up time of demand d
$a_{b\acute{b}}$	Connection state between bus b and $\acute{b}$	$\wp_t$	Auxiliary current time equal to t at time t
		$y_{b\acute{b}}$	Tie-line state between bus b and $\acute{b}$
		n_m	The number of buses in section m

2 Introduction

An electric power system includes three sub-systems: power generation, power transmission, and power distribution. The transmission system as a network includes long transmission lines, transformers and outdoor substations, which are mostly located in open wide areas. It would most likely be affected by any extreme weathers or disasters such as ice or dust storms, hurricanes or earthquakes. Although these disruptions are rare, they have high impact on transmission networks resulting in cases of complete blackouts. For instance, Hurricane Sandy caused long and huge outages over 17 states of the United States in 2012. It also brought terrible monetary losses of over $25 billion dollars to businesses affected [1].

Occurrence of blackouts because of shocks is almost inevitable and the repairing process usually takes time to bring the system back to its normal state. Hence, it is required to investigate in reducing the losses by either preventing the outages or enabling the system to be restored fast. In a restoration process, critical loads could be supplied by additional self-starting generation resources that decrease a large portion of loss of load in a power system. In this order, power networks are equipped with self-starting generation units called BS units. These units can contribute into the sectionalization process, to recover the system with a maximum resiliency [2]. The sectionalization method is a build-up approach. In a build-up restoration, some separated sections are made to be restored individually. Then, a quick restoration process can be conducted by restoring the sections in a parallel fashion. The power system restoration also can save time for the repair crew to fix or replace all damaged components. Therefore, the system can be reconfigured and synchronized to return to its normal state [3, 4].

The quality of a sectionalization-based restoration is dependent on the points of disjoints in the network that can make different topologies over the sections. Depending on the sectionalization pattern, restoration time and the amount of satisfied load might be different. Sectionalization can be performed by minimizing the restoration time [5] or the amount of unmet demand [6, 7]. It is also possible to find an optimal solution with minimum "electrical distance" within each section [8]. From the tie-lines' perspective, minimizing the number of tie-lines could developmore robust sections and consequently a successful restoration [9]. A GPP approach divides a network into a given number of partitions while minimizing the disjoint edges [10]. Therefore, considering the power system network as a connected graph, its buses could represent the graph's nodes and the transmission lines could be equivalent to the graph's edges. Consequently, a joint model of PTSS and GPP can be presented to find the optimal parallel restoration solution.

In this study, a mathematical model is proposed to minimize multiple objective terms including total MWh load shedding, total restoration time and total number of disjoint edges. Furthermore, the power generation cost term is added to the objective to find a restoration plan at lower cost. In order to solve the multi-objective model, preemptive goal programming (PEGP) is used to support optimization of all objective terms at the same time. The model is subjected to the constraints of PTSS and GPP. The PTSS model includes the optimal power flow (OPF) model constraints to perform the restoration as well as sectionalization.

In order to reduce the complexity of the model, a decomposition is applied to divide the model into two levels: the upper level or sectionalization model and the lower level, which is called restoration model. An iterative optimization algorithm (IOA)is used beside PEGP to solve the model. This study brings following major contributions to the literature:

1. Merging the PTSS and GPP models to sectionalize a de-energized power grid.
2. Minimizing the total number of tie-linesconsideringthe place of BS units.
3. Restoring most of the grid's loads within the first hours of a restoration process.

The rest of the paper is structured as follows: Sect. 3 presents the mathematical optimization model. The solution methodology is explained in Sect. 4. Section 5 shows the numerical results and the conclusion is performed in Sect. 6.

3 Model Description

The GPP model minimizes the disjoint edges to split a graph into a given number of partitions as depicted in Fig. 1. The size of each partition could be given as an input or could be found optimally within the model solution. Hence, by combining PTSS and GPP the proposed model will optimize the section sizes while considering the capacity of power generation and minimization of load shedding in each section. Furthermore, cost of restoration time and power generation are minimized to achieve a fast and economic restoration. These objectives are subjected to GPP constraints and PTSS constraints. The GPP mathematical model individually is presented as follows:

$$\min f^{GPP} = \sum\nolimits_{b'=1}^{NB} \sum\nolimits_{b=1}^{NB} y_{bb'} \cdot a_{bb'} \,. \tag{1}$$

$$y_{bb'} = 1 - \sum\nolimits_{m=1}^{NS} s_{mb} \cdot s_{mb'}, \; \forall b, b'. \tag{2}$$

$$\sum\nolimits_{m=1}^{NS} s_{mb} = 1, \;\; \forall b. \tag{3}$$

$$\sum\nolimits_{b=1}^{NB} s_{mb} = n_m, \; \forall m. \tag{4}$$

$$\sum\nolimits_{m=1}^{NS} n_m = NB. \tag{5}$$

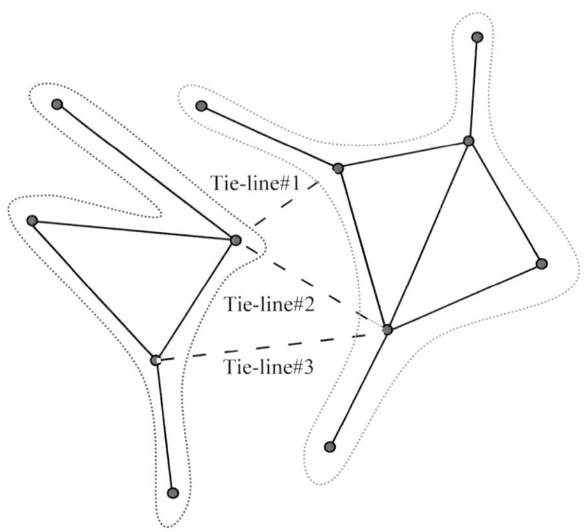

Fig. 1. Graph partitioning example.

The PTSS model is constrained with upper and lower limits of power generation capacity (6), ramp up and ramp down limits of generation units (7).Transmission line power flows constraints are also presented with Eqs. (8)–(10). Since there is no power flow on tie-lines,$y_{b\acute{b}}$ is considered in both (8) and (9) to ensure it. In this study, we set all buses with BS generation units as the reference buses in each section (10). The real power balance at each bus is given in equality constraint (11).

$$P_g^{G,\min} \le P_{gt}^G \le P_g^{G,\max}, \; \forall g, \forall t. \tag{6}$$

$$-RD_g^G \le P_{gt}^G - P_{g(t-1)}^G \le RU_g^G, \; \forall g, \forall t. \tag{7}$$

$$\left|P_{lt}^{L}\right| \leq P_{l}^{L,\max}(1 - y_{bb'}), \forall l \sim (b, b'), \forall t. \tag{8}$$

$$\left|P_{lt}^{L} - \frac{(\theta_{bt} - \theta_{b't})}{x_l}\right| \leq M \cdot y_{bb'}, \forall l \sim (b, b'), \forall t. \tag{9}$$

$$\theta_{ref,t} = 0, \forall t. \tag{10}$$

$$\mathbf{KG} \cdot \mathbf{P}^{G} - \mathbf{KL} \cdot \mathbf{P}^{L} + \mathbf{KD} \cdot \mathbf{LS} = \mathbf{KD} \cdot \mathbf{D}. \tag{11}$$

Furthermore, it is desired to minimize the total load shedding cost, f^{LS}, cost of restoration time, f^{T}, and total power generation cost, f^{Gen}. Therefore, the summation of these three terms is minimized in the objective of PTSS:

$$\min f^{LS} + f^{T} + f^{Gen} = VOLL \cdot \mathbf{LS} + \boldsymbol{\omega}^{T} \cdot (\mathbf{T} \circ \mathbf{s}) + \mathbf{c}^{T} \cdot \mathbf{P}^{G}. \tag{12}$$

Where *VOLL* is the value of loss of load which is equal to \$1000/MWh, $\boldsymbol{\omega}$ is the outage cost vector (\$/h) and is predefined for each class of demand, and $\mathbf{c}$ is the marginal cost of power generation (\$/MWh). There is a Hadamard product ($\circ$) between $\mathbf{T}$ and $\mathbf{s}$ which means an element-wise multiplication of these two same-size matrices [11]. The matrix $\mathbf{T}$ is a given restoration time which is approximated with the minimum possible operational delay to restore a load within each section and load pick-up time of that load which is explained in detail in Sect. 4.

To combine these two models and solve as a linear model, it is required to linearize Eq. (2) by replacing it with the following two inequalities:

$$-y_{bb'} - s_{mb} + s_{mb'} \leq 0, \ \forall (b, b') \sim l, \ \forall m. \tag{13}$$

$$-y_{bb'} + s_{mb} - s_{mb'} \leq 0, \ \forall (b, b') \sim l, \ \forall m. \tag{14}$$

In order to solve the model in a lower complexity, the combined GPP-PTSS model is recast as a bi-level programming.

3.1 Bi-level Programming

The GPP-PTSS model in a bi-level programming structure includes two models: upper and lower levels. The sectionalization would be done through the upper level, which is called "sectionalization" model. The sectionalization model includes GPP constraints as explained and for the sake of simplicity, the PTSS constraints are shown without the line power flow constraints. Therefore, GPP's constraints take care of the power system's network structure and the optimal power flow would be analyzed at lower level, which is called "restoration" model. In the restoration level, the sections' pattern is fixed and the optimal values of load shedding and restoration time could be obtained.

To exclude the transmission line's power flow from the sectionalization level it is required to revise the load balance Eq. (11). Without the line power flow, it is enough to find the load balance per section hence, the network's line power flow limits could be replaced by the total load balance constraint at each section:

$$\mathbf{s} \cdot \mathbf{P}^G + \mathbf{LS}' = \mathbf{s} \cdot \mathbf{D}. \tag{15}$$

Where the first term in (15) is linearized by the given Lemma in [12].

Consequently, the total load shedding cost, f^{LS}, needs to be modified by replacing **LS** (load shedding per load) with $\acute{\mathbf{LS}}$ (load shedding per section), and replacing *VOLL* with $\frac{NB \cdot VOLL}{NS}$. Finally, the upper level minimizes the modified PTSS objective function and f^{GPP} with respect to linearized GPP's constraints and constraints (6), (7), and (11).

The restoration model is constrained by (6)–(11) to minimize the objective function (12). One of the main tasks of the restoration model is to find the optimal restoration time by finding the optimal period taken by BS units to supply the demand, which is called "load pick-up time" [13]. Constraint (16) is in charge of this task while in the objective function f^T is modified to $-\boldsymbol{\omega}^T \cdot \mathbf{T}^{Load}$. The new f^T is negative due to the fact that constraint (16) will find the earliest time to have the zero-load shedding value. Therefore, to avoid $T_d^{Load} = 0$, T_d^{Load} should be maximized in the objective function, which is equivalent to minimizing the negative value of it.

$$T_d^{Load} \leq \wp_t + \mathbf{M} \cdot LS_{dt}, \forall d, \forall t. \tag{16}$$

In the next section, the taken solution methodology to solve the proposed model is described.

4 Solution Methodology

A typical bi-level program can be solved via different approaches such as iterative optimization algorithm (IOA), mathematical programming with equilibrium constraints (MPEC), and penalty function methods. In this study, the IOA method has been applied and the result has been compared to the MPEC approach [14]. In addition, the PEGP method is taken to solve the model at each iteration by considering the multiple objective terms with different scales and priorities.

4.1 Iterative Optimization Algorithm

Iterative optimization method on a bi-level programming is a heuristic algorithm, which solves each level individually and sends updated variables under its control to the other level. The details of the method is illustrated by the flowchart in Fig. 2. As explained in Sect. 3, restoration time is composed of a constant and a variable terms. The constant term contains all delays including switching time, operators handling time, and unexpected delays that are assumed to be given [5]. The variable term is the load pick-up time (T_d^{Load}) which is under control of the lower level model (restoration model) and it is initiated with $T_d^{Load} = 0$. Therefore, by the initial restoration time, the

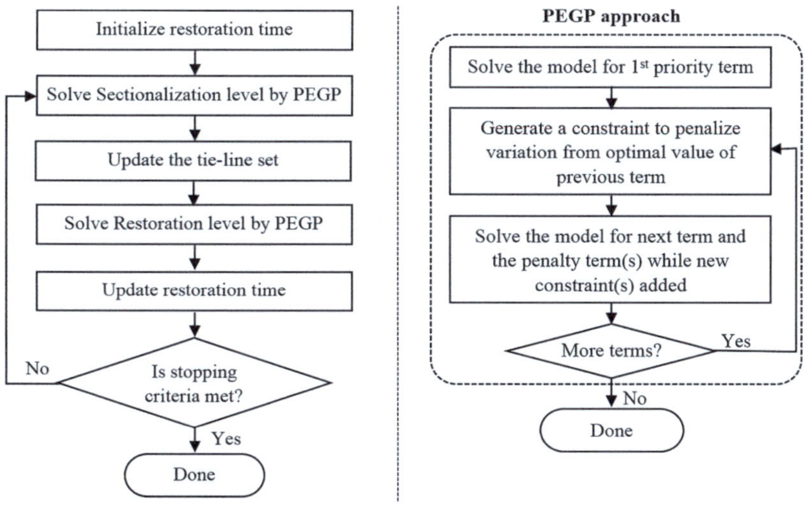

Fig. 2. Iterative optimization algorithm supported by preemptive goal programming

IOA method can start from the upper level (sectionalization model) and iteratively solve the model until stopping criteria get satisfied. Here, the stopping criterion is defined based on the same sectionalization pattern at two consecutive iterations, $\mathbf{s}^k$ and $\mathbf{s}^{k-1}$, as follows:

$$\left\| \mathbf{s}^k - \mathbf{s}^{k-1} \right\|_2 = 0. \tag{17}$$

4.2 Pre-emptive Goal Programming Method

Pre-emptive goal programming divides the goals into different priorities' sets and optimization would be started at the highest priority set, then the next sets are considered such that the optimal value of the previous set is preserved by limiting the feasible area to minimum violation from the optimal value set [15]. In the restoration model, different decision makers come with different priorities in terms of the objective function. In this paper, the importance of each objective term in GPP-based model are chosen as total number of disjoint edges, total load shedding cost, the cost of restoration, and the power generation cost.

At the first step, inequality constraint (18) is the first generated constraint:

$$f^{GPP} \leq f^{GPP*} + \varepsilon^{GPP}. \tag{18}$$

Where f^{GPP^*} is the optimal value of f^{GPP} and ε^{GPP} is the deviation amount from the optimal value, which is penalized at the next objective function:

$$\min f^{LS} + \kappa^{GPP} \cdot \varepsilon^{GPP}. \quad (19)$$

In the new objective function κ^{GPP} is a fixed coefficient for the penalty term to ensure its minimization beside f^{LS}. The same approach is taken at the next steps until particular optimal value of all terms gives the final optimal solution.

The efficiency of the proposed model is examined on two case studies in Sect. 5.

5 Numerical Results

To analyze the performance of the proposed model, the standard IEEE 14- and 118-bus test systems are chosen as small and large-scale test cases. The data regarding these two cases is available at [16]. The small-scale test case assumed to be equipped with three black start (BS) generation units on buses 1, 6, and 10 and the large-scale test case has eight BS units, pre-located on buses 10, 25, 49, 59, 69, 80, 89, and 100. The model is solved with CPLEX 12.3.0.0 under GAMS 24.8.3 on a PC with Intel Xeon 2.53 GHz, 12-core, and 128 GB of RAM. The both cases are solved within the first 24 h after a disturbance. It is also assumed that the systems' post- disturbance statuses are completely de-energized and given as-is.

5.1 Small-Scale Test System Results

Implementing the model on an IEEE 14-bus test system, the optimal sectionalized grid is illustrated by Fig. 3. The sections are well connected and formed by cutting just five lines from 20 lines of the 14-bus grid. The IOA approach converged at the first iteration. The load shedding percentage, restoration time, and line availability percentage are shown in Table 1. Since the IEEE 14-bus test system is not included by the selected comparable study [13], the respected results are produced by the authors of this paper based on the MPEC approach. Table 1 emerges the values of all three investigated elements achieved by GPP-IOA versus MPEC. The main observed reason for the better results by GPP-IOA solution approach is the stronger connectivity, 75% line availability, which causes the higher quality in the restoration process as well.

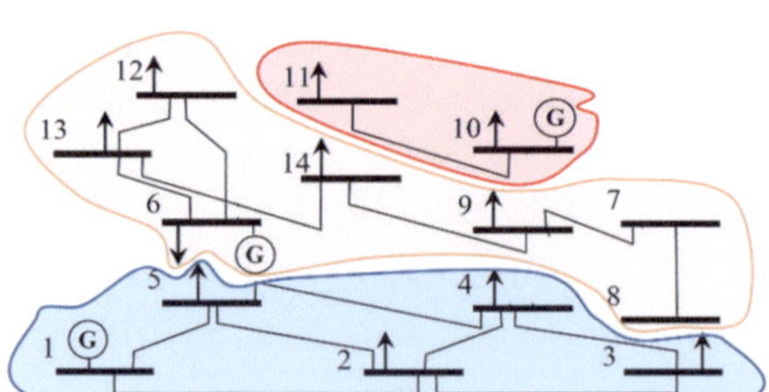

Fig. 3. The sectionalized IEEE 14-bus test system.

Table 1. Comparison between GPP-IOA and MPEC results.

Case	Solution approach	Load shedding %	Restoration time	Line availability%
Small-scale	GPP-IOA	26%	12.66 h	75%
	MPEC	39%	13.11 h	65%
Large-scale	GPP-IOA	18%	10.55 h	83%
	MPEC	26%	–	75%

5.2 Large-Scale Test System Results

The optimal sectionalized grid for IEEE 118-bus test case is presented at Fig. 4.

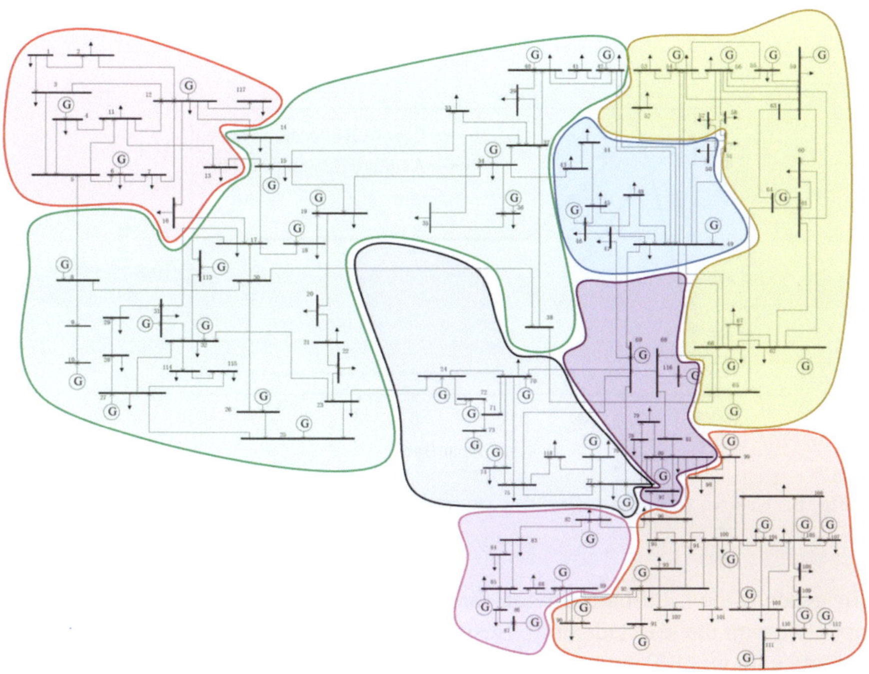

Fig. 4. The sectionalized IEEE 118-bus test system.

Figure 5 shows the total load shedding cost vs the total cost of power generation over iterations. The results as plotted in Fig. 5 shows a drop within the first iterations on load shedding cost while the required power is generated and jumped up as depicted by the cost of generation's curve.

The restoration time is also decreased sharply at the first four iterations as presented by Fig. 6 while a temporary pick value is observed at iteration 2 on the cost of restoration time and both have been stabled since iteration 5 with a restoration time of 10.55 h.

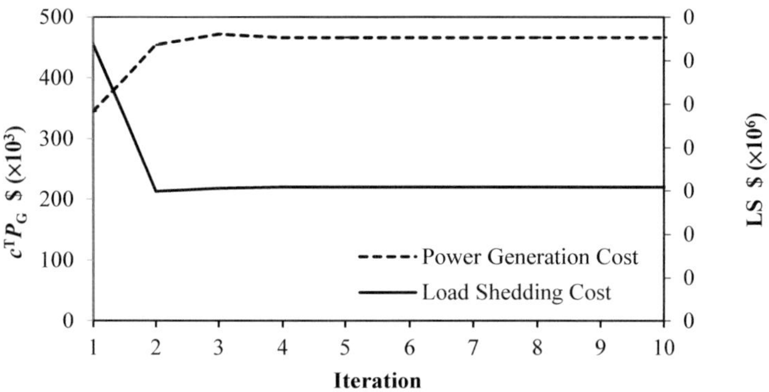

Fig. 5. Total cost of generation vs total load shedding cost on the large-scale case.

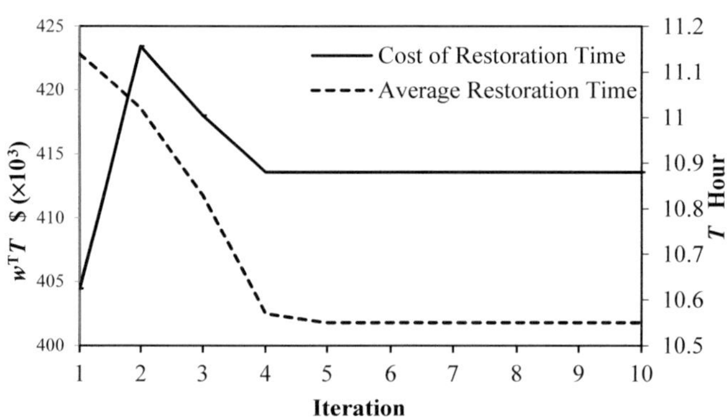

Fig. 6. Cost of restoration time and average restoration time on the large-scale case.

In general, the GPP-based model has a fast convergence and its performance is comparable with the selected study's results solved by MPEC with the same large-scale case study without GPP approach [13]. As presented in Table 1 the GPP model solved by IOA achieved better results in total load shedding value and available line percentages.

6 Conclusion

A power system subjected to a complete blackout must be restored in the lowest restoration time. In a resilient power system, a sectionalization approach would be utilized to have a fast and effective restoration. In this study, a graph partitioning technique is developed to build the sections. The proposed GPP-based PTSS presents a bi-level program solved via IOA and PEGP solution methodologies. The model's efficiency is examined with two case studies: IEEE 14- and 118-bus test systems and the results have been compared with a MPEC method's results from another study. The comparison of the large-scale results shows 8% lower load shedding percentage and 8% higher line availability percentage via IOA solution method in contrast with the MPEC model from the selected study. In conclusion, the presented GPP-based PTSS model develops an efficient optimization framework which gives well-interconnected sections while most of demands are satisfied.

References

1. Blake, E.S., Kimberlain, T.B., Berg, R.J., Cangialosi, J.P., Beven II, J.L.: Tropical cyclone report: hurricane sandy. Nat. Hurricane Cent. **12**, 1–10 (2013)
2. Yutian, L.I.U., Rui, F.A.N., Terzija, V.: Power system restoration: a literature review from 2006 to 2016. J. Mod. Power Syst. Clean Energy **4**(3), 332–341 (2016)
3. Sarmadi, S.A.N., Dobakhshari, A.S., Azizi, S., Ranjbar, A.M.: A sectionalizing method in power system restoration based on WAMS. IEEE Trans. Smart Grid **2**(1), 190–197 (2011)
4. Adibi, M.M., Fink, L.H.: Power system restoration planning. IEEE Trans. Power Syst. **9**(1), 22–28 (1994)
5. Liu, W., Lin, Z., Wen, F., Chung, C.Y., Xue, Y., Ledwich, G.: Sectionalizing strategies for minimizing outage durations of critical loads in parallel power system restoration with bi-level programming. Int. J. Electr. Power Energy Syst. **71**, 327–334 (2015)
6. Liu, W., Lin, Z., Wen, F., Ledwich, G.: A wide area monitoring system based load restoration method. IEEE Trans. Power Syst. **28**(2), 2025–2034 (2013)
7. Abbasi, S., Barati, M., Lim, G.: A parallel sectionalized restoration scheme for resilient smart grid systems. IEEE Trans. Smart Grid (2017, submitted)
8. Cotilla-Sanchez, E., Hines, P.D., Barrows, C., Blumsack, S., Patel, M.: Multi-attribute partitioning of power networks based on electrical distance. IEEE Trans. Power Syst. **28**(4), 4979–4987 (2013)
9. Sydney, A., Scoglio, C., Gruenbacher, D.: Optimizing algebraic connectivity by edge rewiring. Appl. Math. Comput. **219**(10), 5465–5479 (2013)
10. Fan, N., Pardalos, P.M.: Robust optimization of graph partitioning and critical node detection in analyzing networks. In: Wu, W., Daescu, O. (eds.) Combinatorial Optimization and Applications, pp. 170–183. Springer, Heidelberg (2010)
11. Horn, R.A.: The hadamard product. In: Proceedings of Symposia in Applied Mathematics, vol. 40, pp. 87–169 (1990)
12. Glover, F.: Improved linear integer programming formulations of nonlinear integer problems. Manag. Sci. **22**(4), 455–460 (1975)
13. Saadat, H.: Power System Analysis. WCB/McGraw-Hill, New York (1999)
14. Abbasi, S., Barati, S., Lim, G.: A multi-objective MPEC model for disaster management of power system restoration. In: IIE Annual Conference Proceedings. Institute of Industrial and Systems Engineers (IISE) (2017)
15. Ignizio, J.P.: Goal Programming and Extensions. Lexington Books, Lanham (1976)
16. Power Systems Test Case Archive. http://www2.ee.washington.edu/research/pstca

Test and Analysis of Chinese Coal Miners' Vision Ability

Mingming Deng[1,2(✉)] and Feng Wu[1]

[1] School of Management, Xi'an Jiaotong University, Xi'an, China
{dengmm, fengwu}@mail.xjtu.edu.cn
[2] School of Mechanical and Electrical Engineering,
Shaanxi University of Science and Technology, Xi'an, China

Abstract. Two hundred and thirty-nine coal miners' vision, color discrimination, dark adaptation, depth perception, vision response time and Critical Flicker Fusion Frequency (CFF) were tested. The results showed that, with the age increasing, the coal miners' dark adaptation ability, vision response ability and visual sensitivity declined; there were significant differences in vision response ability and visual sensitivity from different types of work; the miners with better vision had stronger dark adaptation ability and visual sensitivity; the miners with stronger dark adaptation ability had stronger visual sensitivity. In the end, some proposals were given to coal miner selection and management.

Keywords: Coal miners · Vision ability · Dark adaptation · Visual sensitivity

1 Introduction

The input and output of human information play important role in the work process. The input of human information mainly depends on human sense system, such as vision, hearing, etc. The output of human information is the response and action according to external stimuli. During the work process, vision is frequently used, which is the main channel for understanding the physical world receiving information [1]. About 80% of the information is obtained through the visual channel [2].

Coal miners work underground for long time. They are often in dark environment, and their vision field is limited. On the one hand they need stronger vision ability, on the other hand the harsh environment of coal mine will do harm to the miners' vision. The older miners' visual function usually decline, such as dark adaptation and color discrimination [3]. In typical physical examination, visual ability is measured by the vision and color discrimination. The vision is measured with far distance naked eyesight, and color discrimination is measured by identifying the main color (such as red, yellow and green, etc.).

Far distance naked eyesight is an important measure in professional selection, such as pilots, military and so on. Color discrimination is an important indicator for professionals in transportation industry [4].

Dark adaptation is the ability of the eye recovering its sensitivity in the dark after exposure to bright lights [2] and it is usually measured by a Dark Adaptation Tester [4]. Kowalski-Trakoxer et al. analyzed that in general older people suffered a loss of contrast

P. Fechtelkotter and M. Legatt (eds.), *Advances in Human Factors in Energy: Oil, Gas, Nuclear and Electric Power Industries*, Advances in Intelligent Systems and Computing 599, DOI 10.1007/978-3-319-60204-2_3

sensitivity, decreases in dark adaptation ability and declines in color sensitivity [3]. Zhang et al. studied technical test of dark adaptation for drivers and showed significant difference of dark adaptation ability between accident and non-accident group [4].

Depth perception refers to one's ability to determine relative and absolute distances within a visual scene [2]. Depth perception can be measured by Depth Perception Tester. One having weaker depth perception ability has weaker spatial perception and stereo perception. Guo et al. tested truck drivers's depth perception ability and found that drivers with weaker depth perception ability were prone to suffer traffic accidents [5].

The response time mainly consists of vision response time and auditory response time. Depending on how much stimulus presentation, the response time can be divided into simple response time and choice response time. Simple response time means response time based on a single stimulus presentation; choice response time means response time based on two or more of the stimulus presentation [2]. Mihal et al. found the choice response time of driver was significantly associated with the accident rate, while simple response time was not. An et al. tested 134 effective participants in a mining company and their results showed that the miners with poor spatial perception and weak response ability performed unsafer [7].

Critical Flicker Fusion Frequency (CFF) is a measure of visually sensitivity. The higher the CFF is, the higher visually sensitivity is. Cui et al. analyzed the change of pilots' cognitive ability by indicators of CFF and vigilance [8].

Existed studies on vision ability of professionals mainly focused on traffic industry, and their test indicators are not comprehensive, not systematic [4–6]. In this paper, the vision, color discrimination, dark adaptation, depth perception, vision response time and CFF of coal mine workers from a coal mine in China were measured and analyzed. Base on the findings, some ergonomics interventions were proposed for coal miner selection and management in China.

2 Method

2.1 Participants

239 coal miners participated in this test study. Each of them was given a souvenir for the participation. Before the test, the participants needed to provide their basic demographic information questions, including age and type of work. There were four response levels for the question on 'age'. 1 meant age <= 30; 2 meant 30 < age <= 40; 3 meant 40 < age <= 50; 4 meant age > 50. On the question on "type of work", there were nine items for choice including digging miner, gas detection miner, carriage miner, coal mining miner, water detection miner, blasting miner assistant, blasting miner, electromechanical miner and miner of other work types. In this study, 237 valid samples were collected, and the valid rate of the test was 99.2%.

2.2 Instruments

Standard logarithmic vision chart, color discrimination chart, Dark Adaptation Tester, Depth Perception Tester, Response Time Tester, and CFF tester, were used for measuring participants' vision, color discrimination, dark adaptation ability, depth perception ability, vision response ability and vision sensitivity of the participants, respectively.

2.3 Procedures

Firstly the participants were told that the tests did no harm to their bodies and health and the process was shown. The tests were conducted in a fixed sequence for all the participants, in sequential order of Vision and color discrimination test, dark adaptation test, depth perception test, vision response time test, and CFF test. Between the tests, the participant was given a rest for one minute.

Vision Test and Color Discrimination Test. Standard logarithmic vision chart was used to measure far distance naked eyesight of the participants in proper environment lighting. The data of vision ranges 4.0 to 5.3. The average value of both eyes was recorder for analysis. Color discrimination chart was applied for participants. 1 means good color discrimination, while, 0 means poor color discrimination.

Dark Adaptation Test. The task of the participant was to observe and read the numbers shown from the viewing window of the tester. The participant was seated in front of the Dark Adaptation Tester and was exposed to strong light surrounding (the surrounding illuminance level was 2000 lx) for 30 s. The tester showed ten number signs one by one in weak light surrounding (the surrounding illuminance level was 1.5 lx). Each sign was presented for five seconds and the participant was required to read it out. The experimenter counted the dark adaptation index, which was the percentage of inaccuracy according to the comparison of the reported and actual number. The higher the dark adaptation index is, the weaker the dark adaptation ability of the participant is.

Depth Perception Test. In this study, Depth Perception Tester was applied, which can measure the participants' depth perception effectively [5]. The participant was seated at a distance of two meters away from the display window of Depth Perception Tester. From the window, two fixed standard stimuli and one compared stimulus of the tester were observed and judged by the participant. Originally, in vertical orientation, the compared stimulus and two fixed standard stimuli were not in a straight line. The participant operated teleswitches of the tester to move the compared stimulus along a guide rail in horizontal orientation to keep the two standard stimuli and the compared stimulus in a vertical straight line. After the participant finished the operation, the experimenter observed and read the distance error between the compared stimulus and two standard stimuli from scale readings of the tester. Each participant was tested twice and two depth perception error data of the participant was read by the experimenter. The mean of absolute value of the two depth perception error data was calculated and noted. The higher the participant's depth perception error is, the weaker the participant's depth perception ability is.

Vision Response Time Test. Response time tester was used to measure vision choice response time in the study. There were four kinds of visual stimuli including red, yellow, green and blue. Participants gave a response according to the different stimulus. Each participant was tested 20 times. The mean of absolute value of the twenty vision response time data was counted and noted. The higher the vision choice response time is, the weaker the vision response ability is.

CFF Test. CFF tester was used to measure CFF in the study. The mean of absolute value of the two CFF data was counted and noted. The higher the CFF is, the stronger the vision sensitivity is.

3 Results

Only three samples had weak color discrimination, so color discrimination of these miners was good in whole. The tested data was analyzed below, except for color discrimination value.

3.1 The Comparison of Vision, Dark Adaptation, Depth Perception, Vision Response Time and CFF of the Coal Miners from Different Age Groups

The tested data was analyzed with Analysis of Variance (ANOVA). The mean comparisons of vision, dark adaptation, depth perception, vision response time and CFF of the coal miners from different age groups were shown in Table 1. There were significant differences in dark adaptation index ($p = 0.000$), vision response time ($p = 0.000$) and CFF ($p = 0.001$) between the coal mine workers from different age groups. It may be seen from the increase in dark adaptation index values with age that dark adaptation ability of the coal mine workers decreased with increasing age. Vision response time increased with age showing that vision response ability of the coal mine workers decreased with age. CFF decreased with age showing that visual sensitivity of the coal mine workers decreased with age.

Table 1. Compare means and Std. deviations of the coal miners from different age groups

Age groups	Vision	Dark adaptation index	Depth perception error	Vision response time	CFF
1 Age < = 30	4.94	2.54	1.57	690.62	34.41
(N = 26)	(0.27)	(1.99)	(1.16)	(131.24)	(1.97)
2 30 < Age <= 40	4.99	3.17	1.80	748.10	32.81
(N = 93)	(0.22)	(2.09)	(2.30)	(163.48)	(2.00)
3 40 < Age <= 50	4.96	3.93	2.35	806.97	32.30
(N = 109)	(0.23)	(2.34)	(2.91)	(138.05)	(2.84)
4 Age > 50 (N = 9)	4.90	6.56	3.09	909.89	32.65
	(0.25)	(2.19)	(2.82)	(109.02)	(1.70)
p	0.515	0.000***	0.187	0.000***	0.001**

Note: N = 237; *** $p < 0.001$, ** $p < 0.01$

3.2 The Comparison of Vision, Dark Adaptation, Depth Perception, Vision Response Time and CFF of the Coal Miners from Different Types of Work

The vision, dark adaptation, depth perception, vision response time and CFF of the coal mine workers doing different types of work was analyzed using Analysis of Variance (ANOVA) and the results are shown in Table 2.

Table 2. Analysis of variance results of vision, dark adaptation, depth perception, vision response time and CFF of coal miners from different types of work

Types of work	Vision	Dark adaptation index	Depth perception error	Vision response time	CFF
Digging miner	4.99 ± 0.18	3.49 ± 2.12	1.81 ± 1.44	809.54 ± 180.31	32.30 ± 2.80
Gas detection miner	4.86 ± 0.32	4.38 ± 2.50	2.65 ± 2.90	821.31 ± 101.26	32.99 ± 1.39
Carriage miner	4.99 ± 0.25	3.91 ± 2.33	1.60 ± 1.35	731.43 ± 134.60	32.84 ± 2.42
Coal mining miner	4.96 ± 0.25	3.46 ± 2.36	2.89 ± 4.18	766.41 ± 167.83	32.47 ± 2.45
Water detection miner	5.05 ± 0.07	2.00 ± 1.41	1.05 ± 0.64	642.00 ± 55.15	31.47 ± 3.49
Blasting miner assistant	5.03 ± 0.17	2.25 ± 1.26	2.73 ± 2.39	800.25 ± 69.53	30.98 ± 2.37
Blasting miner	5.01 ± 0.11	3.14 ± 1.57	1.08 ± 0.81	868.29 ± 179.16	31.83 ± 2.18
Electromechanical miner	4.92 ± 0.26	3.60 ± 2.28	2.51 ± 3.12	719.52 ± 116.74	33.84 ± 1.80
Miner of other work types	4.99 ± 0.22	3.70 ± 2.79	1.39 ± 0.87	803.05 ± 140.07	32.42 ± 2.70
p	0.609	0.779	0.156	0.018*	0.032*

Note: N = 237; * $p < 0.05$

There were significant differences (p = 0.018) in vision response time between the coal mine workers doing different types of work. The highest vision response ability was for water detection miner, followed, in order of declining level of vision response ability, by electro-mechanical miner, carriage miner, coal mining miner, blasting miner assistant, miner of other work types, digging miner, gas detection miner and blasting miner.

There were significant differences (p = 0.032) in CFF between the coal mine workers doing different types of work. The highest visual sensitivity was for electro-mechanical miner, followed, in order of declining level of visual sensitivity, by gas detection miner, carriage miner, coal mining miner, miner of other work types, digging miner, blasting miner, water detection miner, and blasting miner assistant.

3.3 Correlation Amongst Vision, Dark Adaptation, Depth Perception, Vision Response Time and CFF

The correlations among the data of vision, dark adaptation, depth perception, vision response time and CFF were analyzed using SPSS 17, and the results are shown in Table 3. It was found that age correlated significantly with dark adaptation index (r = 0.295, p < 0.01) depth perception error (r = 0.138, p < 0.05), vision response time

Table 3. Correlation amongst age, vision, dark adaptation, depth perception, vision response time and CFF

Factor	Age	Vision	Dark adaptation index	Depth perception error	Vision response time	CFF
Age	1					
Vision	−0.043	1				
Dark adaptation index	0.295**	−0.595**	1			
Depth perception error	0.138*	−0.266**	0.265**	1		
Vision response time	0.304**	−0.07	0.126	0.017	1	
CFF	−0.217**	0.162**	−0.227**	−0.001	−0.166*	1

Note: Values are Pearson's correlation. N = 237; * $p < 0.05$; ** $p < 0.01$.

($r = 0.304$, $p < 0.01$) and CFF($r = -0.217$, $p < 0.05$). Vision correlated significantly with dark adaptation index ($r = -0.595$, $p < 0.01$), depth perception error ($r = -0.266$, $p < 0.01$) and CFF ($r = 0.162$, $p < 0.01$). CFF correlated significantly with dark adaptation index ($r = -0.227$, $p < 0.01$) and vision response time ($r = -0.166$, $p < 0.05$).

4 Discussion

4.1 Chinese Coal Mine Workers' Vision Ability

For the Chinese coal mine workers tested here, dark adaptation ability, vision response ability and visual sensitivity of mine worker declined with increasing age. This was similar with Kowalski-Trakoxer et al.'s study [3], which indicated that the vision of older miners declined. Color discrimination of these miners was good in whole. This was different from Kowalski-Trakoxer et al.'s study [3], which concluded that the color discrimination of older miners declined.

There were significant differences in vision response ability and visual sensitivity for coal miners workers doing different types of work. The type of work with the weakest vision response ability was blasting miner, while the type of work with the highest vision response ability was water detection miner. The type of work with the weakest visual sensitivity was electro-mechanical miner, while the type of work with the highest visual sensitivity was blasting miner assistant. It seems likely that different types of work or work tasks affected vision response ability and visual sensitivity of the mine workers because some type of work required higher vision response ability and visual sensitivity than other types of work.

The coal miners with better vision had stronger dark adaptation ability, stronger depth perception ability and stronger vision sensitivity. From physiological perspective, the indicators of vision ability (such as dark adaptation ability, depth perception ability, vision sensitivity) was related with one another. So the vision of coal miners is an important indicator to measure their vision ability.

The coal miners with stronger dark adaptation ability had stronger depth perception ability and stronger vision sensitivity. Coal miners worked in relative dark surroundings, so the dark adaptation ability of coal miners should be paid with serious attention.

4.2 Ergonomics Proposals

Vision ability, such as dark adaption ability, may have relationship with accident proneness of coal miners [9]. Vision ability test should be well considered in time of coal miners' selection. The dark adaptation ability test and vision test should be taken in coal miners' selection. Dark adaptation ability test has been widely explored in driver selection [4]. It can be applied in coal mine industry.

Regular physical examination including vision ability test is needed for the coal miners. The dark adaptation ability is important for coal miners. The frequency of physical examination including dark adaptation test and vision test should be raised for the older coal miners [10]. The type of work for the older coal miners should be adjusted appropriately.

5 Conclusion

The present study is the first study to explore the vision ability of coal miners systematically and comprehensively. The findings of this research could serve as a useful reference for the government and related organizations when formulating the policies and strategies to help the coal mine improve management. Different high-risk industries can learn from each other in safety management [11]. The present study may give some inspirations to some other industries.

Acknowledgments. This work is supported by grants from the National Natural Science Foundation of China (71471144, 71071126), the Postdoctoral International Exchange Program, and Doctoral Research Start Foundation of Shaanxi University of Science and Technology (BJ15-36). There are no competing interests.

References

1. Mark, S.S., Ernest, J.M.: Human Factors in Engineering and Design, 7th edn. McGraw-Hill, New York (2002)
2. Sun, L.Y.: Human Factor. High Education Press, Beijing (2008)
3. Kowalski-Trakofler, K.M., Steiner, L.J., Schwerha, D.J.: Safety considerations for the aging workforce. Saf. Sci. **43**, 779–793 (2005)
4. Zhang, D.Y., Dai, M.S., Jin, J.: Analysis of dark adaptation for drivers with safety driving at night. China J. Highw. Transp. **12**(4), 107–109 (1999)
5. Guo, W.L., He, C.D., Ge, X.H., Zhou, X.J.: A study of depth perception of truck drivers. Chin. Ergon. **5**(3), 10–13 (1999)
6. Mihal, W.L., Barrett, G.U.: Individual differences in perceptual information processing and their relation to automobile accident involvement. J. Appl. Psychol. **6**(2), 229–233 (1976)

7. An, Y., Zhang, J.S.H., Shao, C.H.B.: Test and research on miners' capacity of safety behaviors. China Saf. Sci. J. **21**(8), 123–128 (2011)
8. Cui, L., Xu, X.R., Wang, L., Wang, L.J., Fu, Z.H.J., Zheng, J., Zhang, S.H., Cui, C.J.: Effect of obstructive sleep apnea/hypopnea syndrome on the cognitive function of pilot. Med. J. Chin. People's Liberation Army **32**(11), 1192–1194 (2007)
9. Deng, M.M., Sun, L.Y., Sun, L.H., Cui, K.: The test of mineworkers' vision response time, dark adaption, audition response time and accident proneness survey. Ind. Eng. Manag. **18** (3), 102–106 (2013)
10. Schieber, F., Fozard, J.L., Gordon-Salant, S., Weiffenbach, J.M.: Optimizing sensation and perception in older adults. Int. J. Ind. Ergon. **7**(2), 133–162 (1991)
11. Grote, G.: Safety management in different high-risk domains—all the same? Saf. Sci. **50**(10), 1983–1992 (2012)

Indonesia's Challenges and Opportunities: Youth Petroleum Engineer's Role to Develop Economic Growth and Natural Resources

Prayang Sunny Yulia(✉), Sugiatmo Kasmungin, and Bayu Satiyawira

Petroleum Engineering Department, Trisakti University, Jakarta, Indonesia
prayang@trisakti.ac.id

Abstract. The situation of global change, competition and decreasing crude oil price have affected many aspects of the petroleum industry in Indonesia and its economic growth. One of the aspects is energy, where there is the choice of oil and gas, or petroleum. Petroleum is still have the biggest demand for the country's development. Meanwhile, in the past few years, there is the reduction of the crude oil prices, which can be both an advantage and disadvantage. Beginning with disadvantage, it is literally affecting human resources, in this case, the petroleum industry's workers. Many workers in the petroleum industry were cut off from their jobs. The decreasing oil price mainly affects oil fields exploration projects, which are high risk, high cost and relies on skilled man-power. One of the possible solutions to overcome current challenges is entrepreneurship. Entrepreneurs can make significant economic growth for the country and can help people retain their jobs. This paper discussed how innovations in the petroleum industry can help growth and prosperity in Indonesia. Indonesia possesses a variety of renewable energy resources, including geothermal, solar, micro-hydro, wind and bio-energy. Alternative fuels have been researched and developed as renewable energy. But it still has challenges, either with the cost, land and regulation. As a result of this analysis, there will be knowledge sharing for many solutions (for example, being an entrepreneur of renewable energy).

Keywords: Indonesia · Youth · Petroleum · Energy · Develop · Sustainable

1 Introduction

The situation of Indonesia's economic growth has been influenced by its natural resources as one of the most important aspects, specifically, fossil energy or petroleum. Petroleum as the fundamental energy for daily and industrial use with high demand. In 2015, the world oil prices have fallen sharply as the figure shown below (Fig. 1):

From the figure below, in the beginning of 2015, crude oil prices have fallen from more than \$100/barrels to \$50/barrels. Things that affecting the fallen of oil prices are weak demand in many countries due to insipid economic growth, coupled with surging US production. US oil production levels were at their highest in almost 30 years. It has been this growth in US energy production, where gas and oil is extracted from shale

P. Fechtelkotter and M. Legatt (eds.), *Advances in Human Factors in Energy: Oil, Gas, Nuclear and Electric Power Industries*, Advances in Intelligent Systems and Computing 599, DOI 10.1007/978-3-319-60204-2_4

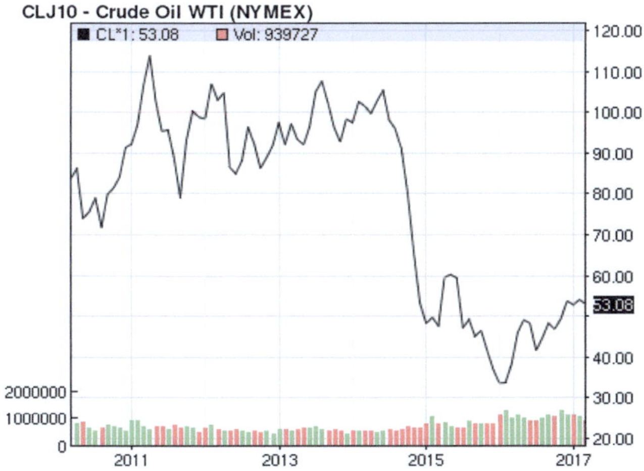

Fig. 1. Crude oil prices graphic from year 2010–2017 [1].

formations using hydraulic fracturing or fracking that has been one of the main drives of lower oil prices [1].

The impact of lower oil prices in Indonesia has its positive and negative sides. The positive sides are the decrease of oil prices will encourage the improvement of Indonesia's current account deficit [2] and give rise of other commodities; for example palm oil and coal. Indonesia is the world's largest producer and exporter of CPO, while the country is also a leading producer and exporter of coal (especially low grade thermal coal) hence collecting valuable foreign exchange through the CPO and coal business. The negative side of the decrease of oil price is, based on Confederation of Prosperous Labor Unions warned that a significant portion of 300,000 employees, contract worker and sub-contractor workers in the oil and gas companies are at risk of being laid off, exacerbating the country's unemployment problem. The unemployment rate in Indonesia reached 6.18% in August 2015, with some 7.6 million people out of Indonesia's 122 million strong workforce out of a job, compared to 7.24 million, or 5.9%, in the same month a year earlier, according to the latest data from the Indonesia's Central Statistics Agency [3].

Facing the uncertainty and volatility condition on economic growth that based on fossil energy or petroleum are the challenges and opportunities for Indonesia. Especially for youth petroleum engineer. Why youth, because based on National Youth Law of Indonesia no. 40 year 2009, youth has 3 (three) main roles for country development, there are; as morality empowerment, social control and agent of change. Then, why petroleum engineer? Because, the main impact is faced by petroleum engineer who directly work on petroleum industries.

2 Indonesia's Economic Growth Overview Based on Natural Resources

The Indonesian economy has recorded strong growth over the past few decades, and in recent years, the firm price of economic expansion accompanied by reduced output volatility and relatively stable inflation. Indonesia's economic performance shaped by government's policy, the country's endowment of natural resources and its young and growing labor force [4].

Indonesia has launched the Law on Energy Number 30 year 2007 to regulate sources, purpose, management of, and policies regarding energy use in Indonesia. Energy by its definition is an ability to produce heat, light, mechanical, chemical and electromagnetic activity (Law of the Republic of Indonesia, 2007). New energy may come from renewable and non-renewable sources, with renewable sources characterized as sustainable and ongoing, including geothermal energy, wind, bio-energy, solar energy, water movement and oceans. Non-renewable sources cannot be sustained and will ultimately run out; they include oil, natural gas, coal and peat. In the national energy mix as of 2006, oil was the main source of energy (52%), followed by natural gas at 29%, coal at 15%, hydropower at 3% and geothermal energy at 1%. Assuming "business as usual" until 2025, the composition of the national energy mix presumes oil to maintain its position as the major energy source at 42%, followed by coal at 35%, natural gas at 21%, hydro power at 1.9%, geothermal energy at 1.1%, and mini/micro hydro power at 0.1% [5].

The indicator of economic growth is from its Gross Domestic Product (GDP) growth. An economy's growth is measured by the change in the volume of its output or in the real incomes of its residents. The 2008 United Nations System of National Accounts (2008 SNA) offers three plausible indicators for calculating growth: the volume of gross domestic product (GDP), real gross domestic income, and real gross national income. The volume of GDP is the sum of value added, measured at constant prices, by households, government, and industries operating in the economy. GDP accounts for all domestic production, regardless of whether the income accrues to domestic or foreign institutions [6]. For Indonesia, the figure below shows on GDP growth for the last 10 years, from 2005–2015 (Fig. 2).

Based on the figure below, the annual GDP growth on the declining oil prices on 2015 is 4.8%, and the graphic shows a declining from 2010. At the following figure, there will be shown total natural resources rent as follows (Fig. 3):

In 2015, the total natural resources rents is 2.324%. According to the two graphics above, natural resources affecting the percentage of GDP. Natural resources give rise to economic rents because they are not produced. For produced goods and services competitive forces expand supply until economic profits are driven to zero, but natural resources in fixed supply often command returns well in excess of their cost of production. Rents from nonrenewable resources - fossil fuels and minerals - as well as rents from overharvesting of forests indicate the liquidation of a country's capital stock. When countries use such rents to support current consumption rather than to invest in new capital to replace what is being used up, they are, in effect, borrowing against their future [7].

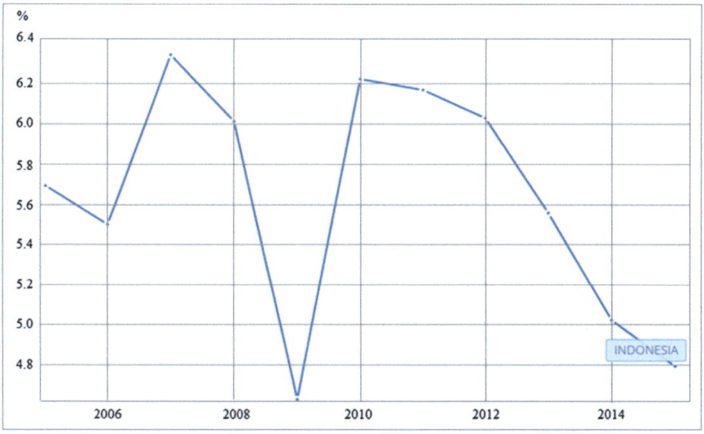

Fig. 2. GDP growth (annual %) [6].

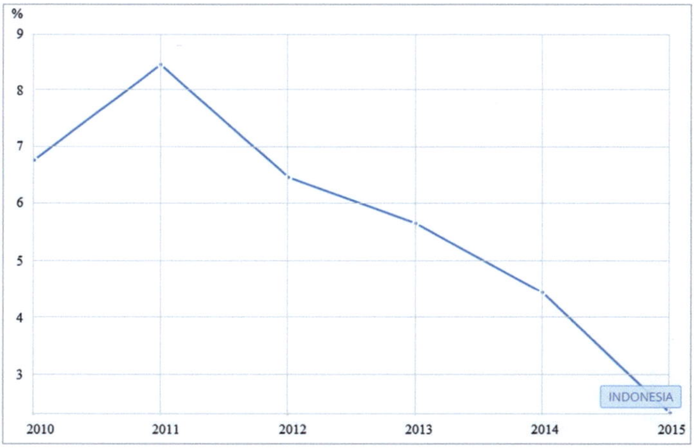

Fig. 3. Total natural resources rents (% of GDP) [7].

Oil price decline had affecting on the beginning of 2015. Based on GDP and total natural resources rents, the graphics shows the decline as well. It assumes that lower oil prices affecting on economic growth, especially on natural resources.

3 Challenges and Opportunities: Youth Petroleum Engineer's Role

With the declining of oil price and its affection to economic growth, it is obvious that the main challenge is unemployment. The lack of investment in petroleum industry, make several oil and gas companies laid off their employees. Rate of unemployment were rising in between year 2014 to 2015, which means the oil price decline was happen on those years. The figure below shown about Indonesia Unemployment Rate, as follows (Fig. 4):

Fig. 4. Indonesia unemployment rate [8].

Young people especially petroleum engineers, have key role for the country development and should take action for implementing a sustainable development. Nowadays, there has an action that called Sustainable Development Goals (SDGs), where it is the continue action after Millennium Development Goals (MDGs). SDGs are the United Nations (UN) initiative that contained 17 goals with 169 targets covering a broad range of sustainable development issues. These included ending poverty and hunger, improving health and education, making cities more sustainable, combating climate change, and protecting oceans and forests (Fig. 5).

All of SDGs goals related to solve the challenges and take the opportunities in order to develop economic growth. However, there are several main goals to reduce

Fig. 5. Sustainable development goals [8]

unemployment. Thing that youth can do to reduce unemployment are being entrepreneur; not as a job seeker yet as a job maker. This thing related with SDGs no. 1 (No Poverty) and SDGs no. 8 (Decent Work and Economic Growth). In Indonesia, entrepreneurship on youth people, especially for petroleum engineer has raised (Fig. 6).

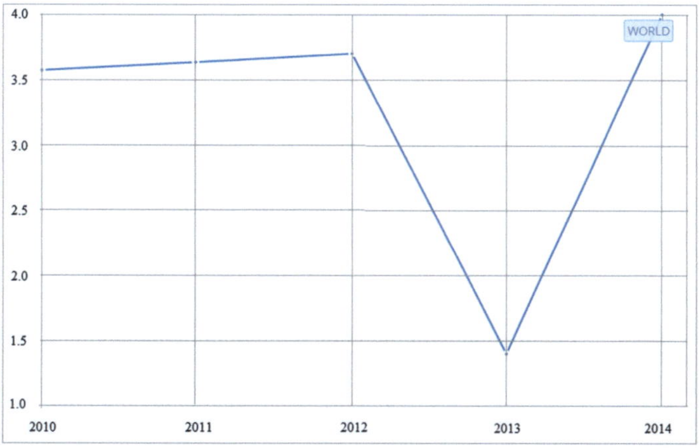

Fig. 6. World bank's entrepreneurship survey and database [9]

The problems and solutions for the affection of oil prices are the unemployment issues and the rise of alternative fuels related to renewable energy. The problem of unemployment in Indonesia, with approximately 700 workers had been cut off from their jobs, thought to be addressed by increasing the number of entrepreneurs. One entrepreneur can help ten people fight poverty. The whole explanation of the problem and solution development in Indonesia is very important to be fixed in order to support sustainable development in Indonesia, especially to help in economic growth. For young generation, it is a mandatory to be aware for these global issues and address them to overcome economic challenges in Indonesia.

Acknowledgements. This research was supported by Faculty of Earth Technology and Energy, Petroleum Engineering Department of Trisakti University. The author would like to express special thanks of gratitude to the Dean of Faculty of Earth Technology and Energy Trisakti University, Dr. Ir. Afiat Anugrahadi, MS.; the Head of Petroleum Engineering Department, Ir. Abdul Hamid, MT.; the Head of Graduate Programme of Magister of Petroleum Engineering Trisakti University, Ir. R.S. Trijana Kartoatmodjo, M.Sc., Ph.D., co-author Dr. Sugiatmo Kasmungin and Ir. Bayu Satiyawira, MSi. from Petroleum Engineering Department of Trisakti University who provided insight and expertise that greatly assisted the research. The author likes to show the gratitude to the parents for prayer and mentally support, and all faculty and department colleagues that give their fully support to the author in making this paper.

References

1. Crude Oil. http://www.nasdaq.com/markets/crude-oil.aspx?timeframe=7y
2. Bowler, T.: Falling Oil Prices: Who are the Winners and Losers? http://www.bbc.com/news/business-29643612
3. Notonegoro, K.: Low Global Oil Prices: Positive or Negative for Indonesia. http://www.indonesia-investments.com/news/todays-headlines/low-global-oil-prices-positive-or-negative-for-indonesia/item6414?
4. Elias, S., Noone, C.: The Growth and Development of the Indonesian Economy. https://www.rba.gov.au/publications/bulletin/2011/dec/pdf/bu-1211-4.pdf
5. Elfani, M.: The impact of renewable energy on employment in indonesia. Int. J. Technol. **1**, 47–55 (2011). ISSN 2086-9614
6. GDP Growth (Annual %). http://data.worldbank.org/indicator/NY.GDP.MKTP.KD.ZG?end=2015&start=1961&view=chart
7. Total Natural Resources Rents (% of GDP). http://data.worldbank.org/indicator/NY.GDP.TOTL.RT.ZS?end=2015&locations=ID&start=2010&view=chart
8. Indonesia Unemployment Rate. http://www.tradingeconomics.com/indonesia/unemployment-rate
9. World's Bank Entrepreneurship Survey and Database. http://data.worldbank.org/indicator/IC.BUS.NDNS.ZS?end=2014&start=2010
10. Suryadarma, D., Suryahadi, A., Sumarto, S.: The Measurement and Trends of Unemployment in Indonesia: The Issue of Discouraged Workers. SMERU Working Paper (2005). ISBN 979-3872-11-X
11. Suryadarma, D., Suryahadi, A., Sumarto, S.: Reducing Unemployment in Indonesia: Result from a Growth-Employment Elasticity Model. SMERU Working Paper (2007). ISBN 978-979-3872-33-9
12. Sjöholm, F.: Economic recovery in Indonesia: the challenge of combining FDI and regional development. Department of Economics, National University of Singapore (2000)
13. Foster, I., Kesselman, C., Nick, J., Tuecke, S.: The physiology of the grid: an open grid services architecture for distributed systems integration. Technical report, Global Grid Forum (2002)
14. Yudha, S.W.: Low Oil Prices: Impact for Indonesia. http://nbr.org/research/activity.aspx?id=557#footnote1
15. Low Oil Prices a Risky Boon for Indonesia's Economy. http://jakartaglobe.id/business/low-oil-prices-risky-boon-indonesias-economy/
16. National Youth Law of Indonesia no. 40 year 2009 (Undang-Undang Kepemudaan Republik Indonesia). http://undangundang.net/uu-nomor-40-tahun-2009-tentang-kepemudaan.html
17. Sustainable Development Goals. https://sustainabledevelopment.un.org/sdgs

Associating Holography Techniques with BIM Practices for Electrical Substation Design

Alexandre Cardoso(✉), Isabela Cristina do Santos Peres, Edgard Lamounier, Gerson Lima, Milton Miranda, and Igor Moraes

Faculty of Electrical Engineering, Federal University of Uberlândia, Av. João Naves de Ávila, 1111, Bairro Santa Mônica, Uberlandia, MG, Brazil
alexandre@ufu.br

Abstract. The emerging of BIM (Building Information Modeling) techniques will change traditional procedures of design and maintenance for electric substations. In addition, Computational Holography, supported by wearable computers, has the potential to allow simultaneous engineering work, based on mixed reality and computer vision capabilities. It is believed that this set of tools will increase engineering design decisions. In this work, we propose a set of techniques to support a complete substation design, which is created by using BIM concepts that explore the Holographic world benefices. Experiments have shown that the coupling of these techniques has the potential to reduce the learning curve of the users, since it changes the way of collaboration among different professional specialists considering simulation intents.

Keywords: Computational holography · Building information modeling · Mixed reality · Power electrical substation

1 Introduction

An electric power substation is considered as a highly engineering critical system. Therefore, computer-based simulation tools play a very important role in preparing engineers to deal with their safety issues. In recent years, with the advance of multimedia and virtual reality, the energy sector has been benefited in deriving simulation environments, based upon such techniques.

In fact, these systems allow for reducing maneuver times, facilitating communication between remote control rooms and on-site operators as well as improving the quality of the service [1, 2]. In addition, Mixed Reality (MR) has profited from this progress, both on sophisticated and popular platforms [3]. Unlike Virtual Reality (VR), which transports the user to the virtual environment, Mixed Reality keeps the user in their physical environment and transports a 3D model into the user's space, allowing a natural and intuitive interaction.

Some studies have explored MR for many architectural and construction applications. For example, [4] used MR to overlap sites of utility lines in real world views to demonstrate the potential of MR to help maintenance workers, avoiding buried infrastructure and structural elements [4]. This technology is being improved based on interaction devices to allow more natural hands manipulation of the virtual objects inside the user space.

P. Fechtelkotter and M. Legatt (eds.), *Advances in Human Factors in Energy: Oil, Gas, Nuclear and Electric Power Industries*, Advances in Intelligent Systems and Computing 599, DOI 10.1007/978-3-319-60204-2_5

In the same way, BIM initiatives has grown with design tools that provide capabilities required for a full specification of a complete Substation at the project phase, for example SDS Design Suite® for Inventor® [5].

Actually the Computational Holography term can also be classified as a Mixed Reality technique that allows a complete representation of Virtual Models mixed with the real world provided by wearable devices using computational vision capabilities.

Microsoft HoloLens® provides a set of algorithms to interact with virtual and real information in a very natural way. With gesture and speech commands, it allows one to visualize and work with the engineering project as part of the real world [6]. Thus, the general objective of this work is to demonstrate the feasibility of the association of BIM practices and Mixed Reality application, associated to a Microsoft HoloLens®, for management and planning of electrical substation procedures. Power substation projects are increasingly complex and require shorter delivery times. Besides, the need to reduce costs compels engineers not only to optimize their layouts, but also to use design procedures with the maximum efficiency, without giving up maximum security with respect to the information generated [7].

1.1 BIM

The National Building Information Model Standard Project Committee (NBIMSPC) defines BIM in the following manner [8]: "BIM is a digital representation of the physical and functional characteristics of a facility. It consists of a single, shared source of information about an installation, constituting a reliable database for decisions throughout its lifecycle, from preliminary studies to the end of its useful life".

In this way, BIM brings quantities and properties of components such as technical characteristics, examples and manufacturers, not just a 3D design, that is, a model contained by the geometry of the installations. BIM makes it imperative to represent a project as a combination of "geometric models", their attributes, and the relationship of these "geometric models" with other components. One aspect of BIM software is that they define the objects in parametric terms and from relations with other objects, so that if a parameter of an object is changed, all dependent parameters are automatically updated [7].

BIM offers the opportunity to realize the project in 'virtual' space and to define the necessary information before the use of resources and to commit to the construction of an installation that will exist for twenty years or more. Instead of encountering problems and conflicts during the construction phase, BIM offers the opportunity - to all participants in the design process - to use computer-based applications to, for example, visualize the 3D model, so to anticipate problems while the projects are still in conceptual phase. This makes it possible to identify and select the most appropriate form of action in the correction or improvement of the project, long before the detail, specification or construction of the installation [8].

For professionals involved in a project, BIM allows the virtual model to be transferred from the design team to the construction and assembly teams, including sub-suppliers, and then to the owner. Each professional adds their discipline-specific data to the database, which will be shared with the entire team. This reduces the loss of information that traditionally occurs between different teams involved, and provides more complete information to owners of more complex facilities. With this, the BIM philosophy is useful and should be applied to the substation design process. For the specific case of large electrical substations, the BIM process can be further accelerated if the BIM tool allows the use of standardized components, structures and assemblies, and by the reuse of these sets in the projects [7].

1.2 Mixed Reality/Computational Holography

Second [9] Mixed Reality (MR) can be defined as the "amplification of sensory perception through computational resources". It allows a more natural interface when working with computer-generated data and images across computing data and real-world information.

One of the most important characteristics of MR is that the interaction occurs with the environment that surrounds the user interacting, guaranteeing him the conditions to interact with this data naturally. Thus, MR uses the association of VR and real environment, offering the user a better perception of the environment and favoring their interaction.

Some of the main features of MR are real-time processing and combining real and virtual elements. By such characteristics, the design of Mixed Reality solutions requires components that allow the assessment of the position of the user (user registration), point of view, and generate the virtual elements to finally combine them with the real world, by means of a projection system. Thus, Mixed Reality systems require hardware to capture information from the environment where the user is, software for real-time generation of virtual elements and hardware to map such elements in the real world [3].

Thus, one of the systems of projection, mapping and processing capable of creating augmented reality is HoloLens®. This device is a wearable computer that designs holograms on the environment that the user is. User interaction is quite realistic because it is done through gestures and voice commands. All features of the glasses are inside the device, including CPU, GPU and an "HPU" (Holographic Processing Unit), letting the user move freely in any environment. The holography that HoloLens® produces can be classified as a particularity of MR called mixed reality (MR). A can be defined as the overlapping of computer-generated three-dimensional virtual objects with the physical environment, shown to the user, supported by some real-time graphic computing device [10]. The goal of a Mixed Reality system is to create such a realistic environment that the user does not perceive the difference between the virtual elements and the actual participants of the scene, treating them as one thing.

2 Related Work

AR habilitation technologies have been researched in a growing number of studies in recent years. By presenting contextual information in textual or graphic form, the real-world user's view is increased or increased beyond normal experience. The addition of such spatially localized computer-generated contextualized information to the user can aid in the performance of various scientific and engineering tasks.

Thus, some studies related to the development of applications based on Virtual or Augmented Reality techniques that used the concepts of BIM for design, maintenance or teaching with the development of engineering systems [11] were studied. The Building Information Modeling and Mixed Reality strings were used to define the selected jobs.

1. The Collaborative Augmented Reality Based Modeling Environment for Construction Engineering and Management Education [11]. This paper presents the initial results of a project aimed at the learning process in construction, an interactive Mixed Reality (MR) learning tool was designed and implemented to help students develop a comprehensive understanding of construction equipment, processes, and operational safety. Figure 1 index 1 shows a step of the system.
2. BIM | MAR: Assembling physical objects by virtual information [12]. In this paper, an investigation was made with the objective of creating a prototype tool that could overcome the most common errors in each of the four stages of the assembly process of civil construction structures, designed by BIM, And developed a low cost solution to translate geometry and data from a BIM to an AR environment. Figure 1 index 2 shows a step of the system.
3. Augmented Reality for Substation Automation by Utilizing IEC 61850 Communication [13]. This article combined IEC 61850 with augmented reality to add user-visible information to a power substation. This used markers and a smartphone to implement the system, which connects with SCADA to receive the information. Figure 1 index 3 shows a step of the system.
4. Mobile Cross-Media Visualizations made from Building Information Modeling Data [14]. Six types of demonstrators made from the same data were presented using similar workflows for different output devices. These devices range from head-mounted monitors to smartphones and tablets with positioning tracking from the inside out. They display cross-media previews that ranged from a sophisticated car-based AR-setup, wired and wireless VR, to view through AR in smart glasses and video-based AR on tablets. Figure 1 index 4 shows a step of the system.
5. WebVis BIM: Real Time Web3D Visualization of Big BIM Data [15]. In this work a BIM data organization strategy for interactive WEB3D was developed. It was possible to visualize BIM in real time on the web, and due to the large volume of data the authors realized that it was impossible without a progressive loading. They used tool like bigIFC, incremental Frustum of Interest (FOI) and they divided the products of the exterior interior of the building. Figure 1 index 5 shows a step of the system.
6. Virtual Training of Fire Wardens Through Immersive 3D Environments [16]. This article presented an interactive RV application to train occupational risk prevention specialists and more precisely on fire safety in buildings. The platform allows the simulation of fires throughout the building. Simulated fire will react with environment, propagating itself in different ways depending on several factors. The BIM standards were applied to the logistics of the simulation, giving information about the building. Figure 1 index 6 shows a step of the system.

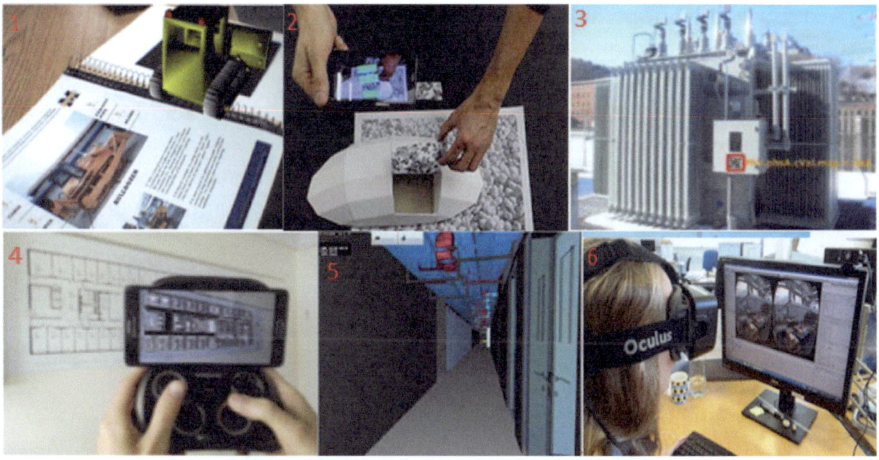

Fig. 1. Image of related works

2.1 Comparative Work

As can be seen in Fig. 1, we have representative images of related works.

Table 1 presents a judgment of the works correlated to the pertinent aspects in the elaboration and use of BIM and RA.

Table 1. Comparative table of related works.

Published works	BIM	AR with using bookmarks	VR	Energy substation
1	X	X		
2	X	X		
3		X		X
4	X	X		
5	X		X	
6	X		X	

With the study, it was possible to verify that, no work associated holography, BIM and electric power substations. So, the goal of this work is to use HoloLens® to synchronize the current location and insert a BIM model into the real environment. This will give each user the ability to go to a workplace and see the future of building within the model, from their own perspective, interacting with the other users. With the wealth of data that BIM provides, the user will can visualize a proposed 3D model in the real environment, even if it is in a space to be remodeled [7].

3 Description of the Developed Environment

The methodology proposed in this paper is composed of the following stages (Fig. 2):

(a) Acquisition of information regarding the features of the substation (CAD plans, photos, videos and equipment catalogues), by means of a standardized protocol;
(b) Definition of techniques to model three-dimensional components of a substation contemplating its constructive and necessary information for the purposes of simulation control and maintenance;
(c) Automatic generation of the three-dimensional environment (automatic positioning of equipment from the library, reuse of electrical arrangements, topology for equipment start-up);
(d) Standardized interface templates for best navigation control, reading of electric component information and command sending.

3.1 Creating and Exporting BIM Information to Unity

Due to the large amount of information contained in a BIM model this becomes a candidate to be imported into a game engine to build a realistic virtual environment. However, this process requires several steps and may vary depending on the use of the environment as shown at Fig. 2.

Fig. 2. Pipeline of the prototype.

From the photos and construction documents (CAD, component files, manufacturer documentation etc.), one initiates the construction of 3D models that will go on to construct the virtual environment (physical and information modeling), in Inventor®. The animation each model is modeled in 3DSMAX®, then is validated and inserted into a Model Library, which further groups together photos and other documents.

Thus, the system allows for the converting of CAD models into virtual environments (semi-automatic generation for virtual environments, by means of the 2D CAD project) [17].

By using such a mechanism, an incomplete VR environment is generated, without cables and connections between the distinct virtual objects. To simplify the process cables and connections was suppressed. In turn, the design is sent forward to the association stage. Here, each element from the virtual model is associated to an identifier within the SCADA system [18], thus generating conditions in the virtual environment, for presenting the state of each monitored component.

3.2 Acquisition Protocol

A site survey at the substation location is necessary for obtaining the wealth of pictures essential to providing the real profile and connection between its elements (e.g. cables, electric bus, etc.). This procedure was used as a cheaper and faster alternative to 3D

Scan, guarantying precision and similarity. The digitalization process is the first challenge [17] when it comes to obtaining the 3D model of each component of the power substations. Hence, one of the main causes of delay encountered in the execution of projects that require 3D modeling is the lack of the necessary documentation for carrying out the entire process. Elements that make up this documentation include CAD plant design, photos and videos with a substation and equipment field survey, plus technical equipment catalogues (data sheets), such as circuit breakers, disconnecting switches, and other substation components, all of which are paper based documentation, due to the fact that many power substation have been in operation for more than 10 years.

3.3 Mixed Reality (MR) Environment

The developed MR environment allows for the construction mode of a complete scenario of the power substation, presented in Fig. 3.

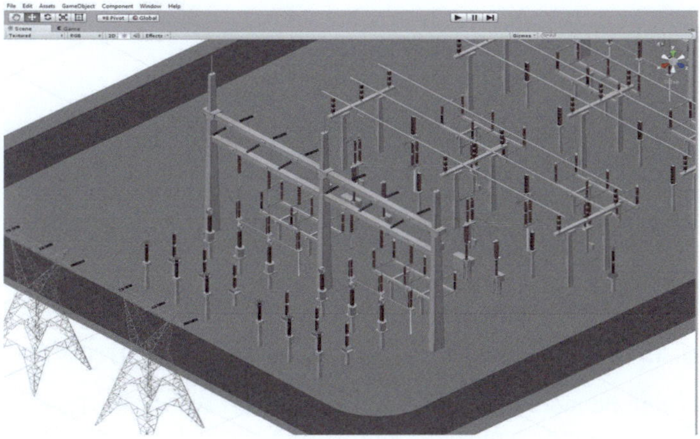

Fig. 3. Complete substation (construction mode)

3.4 Demonstration of the Prototype

In order to support engineer project and maintenance of electrical substation, the use of BIM practices and holography bring to us concurrent engineering (CE). It is systematic approach engineering to the integrated can be defined simultaneous to be design of a product and the related processes, including manufacturing and the other support functions.

Some benefits of CE that [19] studies were producing higher products at lower cost and in less time than previously. Because the quality of designs increases around 50%, development cycle time reduced by as much as 40–60%, manufacturing costs reduced around 30–40%, scrap and rework reduced by as much as 75% and reduced maintainability.

In this context, an interaction between users and system has been developed to facilitate team work, maintenance and administration of the work. Therefore, after the generation of the virtual substation environment, the template containing the full interface description is added to it. This, in turn, will be incorporated across all substations, thus reducing the time spent for adjusting these components.

Through the means of an internal architecture, data referring to the state of the electrical equipment (turned on or off, electrical measurements) is received and processed in real time via WebService. These data are made available by the SCADA system at the Cemig Control Center [17].

The prototype of this work, in Fig. 3, shows the holography created by Hololens®. This is shared to computer screen HMD VIVE® by internal network. The substation seen in Holography, Figs. 4 and 5, is a perfect copy of a substation that is operating in the state of Minas Gerais (CEMIG).

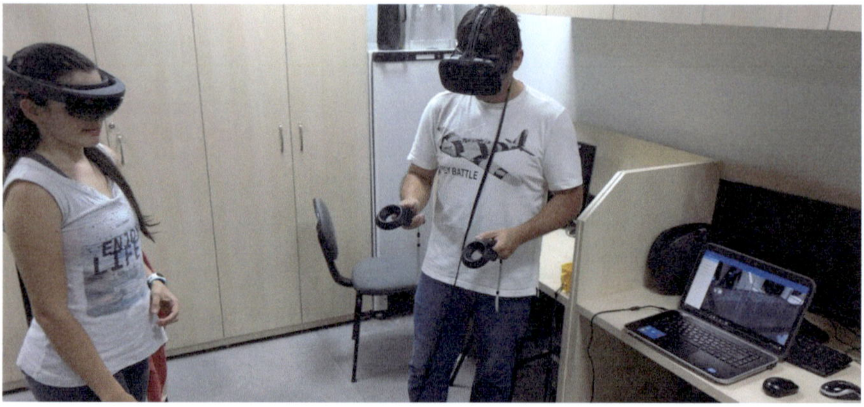

Fig. 4. Interaction between the computer, HoloLens® and HTC VIVE®.

To interaction be more natural we choose don't do any menu. Then, to interact to the system we configure commands with gesture and voice. In Fig. 6 a switch disconnector was selected, with this selection the substation is deactivated. For information about this key it is necessary to say "information" and, while it is selected, by saying "open" it will initiate the opening animation.

This system is integrated with the SCADA (supervisory control and data acquisition) [18] through interface integration. This interface, besides making the connection between the SCADA and the prototype, behaves as a layer of compatibility between the two systems and permitting the continual flow of information related to management. It prevents the direct connection to SCADA, and ensures access to various data sources, such as the names and properties of the elements of substations and power plants, as well as the parameters and measurement values of each element.

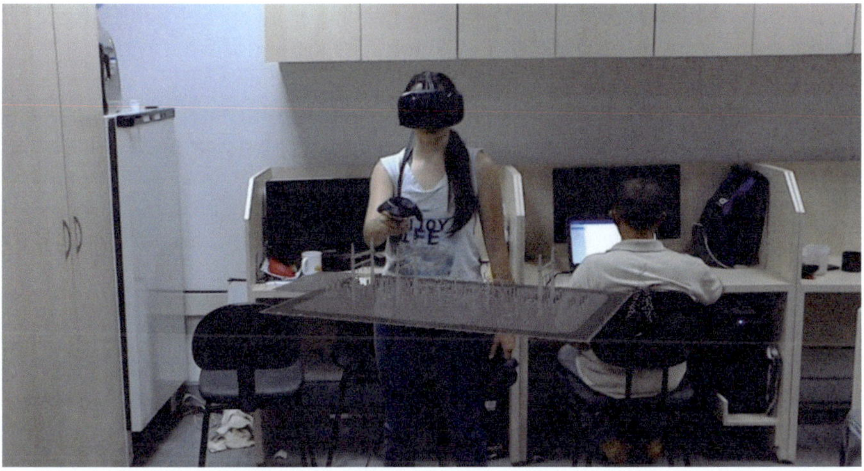

Fig. 5. Holography seen through HoloLens®.

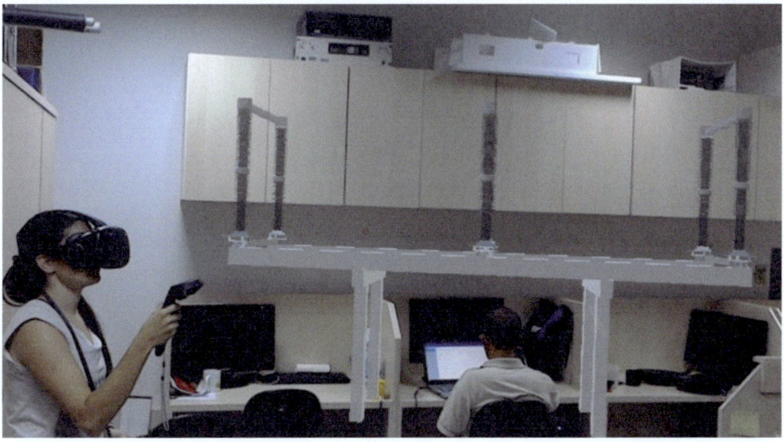

Fig. 6. User interaction with electrical disconnect switch

4 Conclusions

In Mixed Reality, people, places, and objects of their physical and virtual worlds merge into a blended environment that becomes the user's vision. Feeling present in the environment is a key aspect of mixed reality, allowing you to move naturally, interact and explore in three dimensions dynamically. Microsoft HoloLens®, through blended reality, allows you to interact with content and information in the most natural way possible as it allows you to view and work with the project as part of your real world [3, 17].

With it is possible to work close to reality, with the location of the mapped project and geolocation, the project can be viewed in real size, shared works with colleagues and modified in real time. Thus, it makes the decision making smarter and lessens the chance of errors.

A BIM-associated Mixed Reality system enhances the spatial perception of BIM geometric models and contributes to a better understanding of what is being proposed [1, 20]. Thus, performing an initial validation of the BIM and MR, it can be reported that HoloLens® presented a satisfactory performance, even with limited processing. And due to the natural interaction that HoloLens® allows the user an easy cognition of the application [7].

Through the creation and application of the protocol for acquiring information from substations, it was possible to manage in an effective manner the issues related to the data necessary for initiating the construction process, which also made it an effective instrument for the validation of this material.

Regarding the convention rules for modeling, it was possible to identify that these are fundamental to the process that is associated with automation tasks, besides providing improved performance and fluidity while navigating through the system.

For future work we intend to create a large library of models so that the user can modify the environment, implement functionalities such as cost calculation and validate the system in conjunction with a team of specialized designers.

Acknowledgements. The authors would like to thank CEMIG GT411 Research and Development Program, along with CAPES (Coordination for the Improvement of Higher Education Personnel), for all their financial support given to this research project.

References

1. Okapuu-von Veh, A., et al.: Design and operation of a virtual reality operator-training system. IEEE Trans. Power Syst. **11**(3), 1585–1591 (1996)
2. Quintana, J., Mendoza, E.: 3D virtual models applied in power substation projects. In: 15th International Conference on Intelligent System Applications to Power Systems, ISAP 2009, pp. 1–3 (2009)
3. Kirner, C., Kirner, T.: Evolução e Tendências da Realidade Virtual e da Realidade Aumentada. Pré-Simpósio de Realidade Virtual e Aumentada, Uberlândia (2011)
4. Roberts, G., Evans, A., Dodson, A., Denby, B., Cooper, S., Hollands, R.: The use of augmented reality, GPS and INS for subsurface data visualisation. In: Proceedings of the XXII International Congress of the FIT, Washington, D.C. (2002)
5. Substation design. http://substationdesignsuite.com
6. Microsoft Hololens. https://www.microsoft.com/microsoft-hololens/en-us
7. Vasconcellos, A.S., Hernandes, L.: Aplicação de Metodologia BIM no Projeto e Construção de Grandes Subestações. Seminario Nacional de Produção e transmissão de Energia Elétrica, Foz do Iguaçu, PR (2015)
8. The National BIM Standard: Frequently Asked Questions About the National BIM Standard-United States™. http://www.nationalbimstandard.org/faq.php#faq1
9. Cardoso, A., Lamounier, E.: A Realidade Virtual na Educação e Treinamento. In: Fundamentos e Tecnologia de Realidade Virtual e Aumentada. VIII Pré-Simpósio de Realidade Virtual e Aumentada, Belém, pp. 304–312 (2006)
10. Tori, R., Kirner, C.: Fundamentos de Realidade Aumentada. In: Fundamentos e Tecnologia de Realidade Virtual e Aumentada. VIII Pré-Simpósio de Realidade Virtual e Aumentada, Belém, pp. 2–21 (2006)

11. Behzadan, A.H., Iqbal, A., Kamat, V.R.: A collaborative augmented reality based modeling environment for construction engineering and management education. In: Proceedings of the Winter Simulation Conference, Winter Simulation Conference, pp. 3573–3581 (2011)
12. Buyuklieva, B., Kosicki, M.: BIM|MAR: assembling physical objects by virtual information. In: Proceedings of the 4th International Symposium on Pervasive Displays, pp. 257–258. ACM (2015)
13. Antonijević, M.S., Keserica, H.: Augmented reality for substation automation by utilizing IEC 61850 communication. In: MIPRO 2016, 30 May–3 June 2016 (2016)
14. Oppermann, L., Shekow, M., Bicer, D.: Mobile cross-media visualisations made from building information modelling data. In: Proceedings of the 18th International Conference on Human-Computer Interaction with Mobile Devices and Services Adjunct, pp. 823–830. ACM (2016)
15. Liu, X., Xie, N., Jia, J.: WebVis_BIM: real time Web3D visualization of big BIM data. In: Proceedings of the 14th ACM SIGGRAPH International Conference on Virtual Reality Continuum and its Applications in Industry, pp. 43–50. ACM (2015)
16. Diez, H.V., Garca, S., Mujika, A., Moreno, A., Oyarzun, D.: Virtual training of fire wardens through immersive 3D environments. In: Proceedings of the 21st International Conference on Web3D Technology, pp. 43–50. ACM (2016)
17. Cardoso, A., Prado, P.R., Lima, G.F.M., Lamounier, E.: A virtual reality based approach to improve human performance and to minimize safety risks when operating power electric systems. In: Cetiner, S.M., Fechtelkotter, P., Legatt, M. (eds.) Advances in Human Factors in Energy: Oil, Gas, Nuclear and Electric Power Industries, pp. 171–182. Springer, Cham (2017)
18. Boyer, S.A.: SCADA: supervisory control and data acquisition. International Society of Automation (2009)
19. Keys, L.K., Rao, R., Balakrishnan, K.: Concurrent engineering for consumer, industrial products, and government systems. IEEE Trans. Compon. Hybrids Manuf. Technol. **15**(3), 282–287 (1992)
20. Kurabayash, S., Ishiyama, H., Kanai, M.: Sensing-by-overlaying: a practical implementation of a multiplayer mixed-reality gaming system by integrating a dense point cloud and a real-time camera. In: IEEE International Symposium on Multimedia (2016)
21. Bille, R., Smith, S.P., Maund, K., Brewer, G.: Extending building information models into game engines. In: Proceedings of the 2014 Conference on Interactive Entertainment, pp. 22:1–22:8. ACM (2014)
22. Genty, A.: Virtual reality for the construction industry: the CALLISTO-SARI project, benefits for BOUYGUES CONSTRUCTION. In: Proceedings of the 2015 Virtual Reality International Conference, pp. 11:1–11:7. ACM (2015)
23. Hagedorn, B., Döllner, J.: High-level web service for 3D building information visualization and analysis. In: Proceedings of the 15th Annual ACM International Symposium on Advances in Geographic Information Systems, pp. 8:1–8:8. ACM (2007)
24. Moon, H.-S., Kim, H.-S., Moon, S.-Y., Kim, H.-W., Kang, L.-S.: Prototype model of wireless maintenance operation system for bridge facility using VR and ubiquitous technology. In: Proceedings of the 11th ACM SIGGRAPH International Conference on Virtual-Reality Continuum and its Applications in Industry, p. 350. ACM (2012)
25. Ock, J., Issa, R.R.A., Flood, I.: Smart building energy management systems (BEMS) simulation conceptual framework. In: Proceedings of the 2016 Winter Simulation Conference, pp. 3237–3245. IEEE Press (2016)
26. Azuma, R., et al.: Recent advances in augmented reality. IEEE Comput. Graph. Appl. **21**(6), 34–47 (2001)

An Intelligent Operator Support System for Dynamic Positioning

Jurriaan van Diggelen(✉), Hans van den Broek,
Jan Maarten Schraagen, and Jasper van der Waa

TNO, Kampweg 5, 3769 DE Soesterberg, The Netherlands
{jurriaan.vandiggelen,hans.vandenbroek,
jan_maarten.schraagen,jasper.vanderwaa}@tno.nl

Abstract. This paper proposes a human-centered approach to Dynamic Positioning systems which combines multiple technologies in an intelligent operator support system (IOSS). IOSS allows the operator to be roaming and do other tasks in quiet conditions. When conditions become more demanding, the IOSS calls the operator to return to his bridge position. In particular, attention is paid to human factors issues such as trust misalignment, and context-aware interfaces.

Keywords: Cognitive systems engineering · Personal assistants · Dynamic positioning · Predictive analytics

1 Introduction

Dynamic positioning (DP) is a computer-controlled system which aims to maintain a vessel's position and heading using dedicated propellers and thrusters. DP operations form the basis of Floating Production, Storage and Offloading (FPSO) platforms and are a typical example of a highly automated control task that still requires human supervision: four operators are working in shifts 24/7 to monitor the system and resolve malfunctions in the rare case that this cannot be done automatically by the DP system. Sensor values that exceed threshold values lead to alarms, and serve as the primary means to trigger operators to solve malfunctions and abnormalities.

Because the DP operator (DPO) is not busy most of the time, relatively high personnel costs are spent on little work, and the operator could suffer from problems like drowsiness and boredom. A more self-sufficient control system capable of dealing with an increased range of conditions, would not solve this problem by itself. The DP operator would be even less occupied during his work shift, but would still be required to solve 'left over' incidents.

We believe that alarm-based DP systems cannot be advanced further to solve this impasse (which is sometimes referred to as the automation paradox [1]). Therefore, we propose a human-centered approach to DP systems which combines multiple technologies in an intelligent software agent, called IOSS (intelligent operator support system). IOSS functions as a team-partner of the DPO [2] and allows the operator to be roaming and do other tasks in quiet conditions. When conditions become more

P. Fechtelkotter and M. Legatt (eds.), *Advances in Human Factors in Energy: Oil, Gas, Nuclear and Electric Power Industries*, Advances in Intelligent Systems and Computing 599, DOI 10.1007/978-3-319-60204-2_6

demanding, the IOSS calls the operator to return to his station position at the bridge. Ultimately, this could save costs by deploying personnel more efficiently. Furthermore, it creates a more varied job description for DPO's than just system monitoring.

We have followed a systematic approach that integrates technological, human factors (HF), and operational perspectives (i.e. situated Cognitive Engineering (sCE) [3]) to develop the first prototype of IOSS. The four steps in the process are depicted below (Fig. 1).

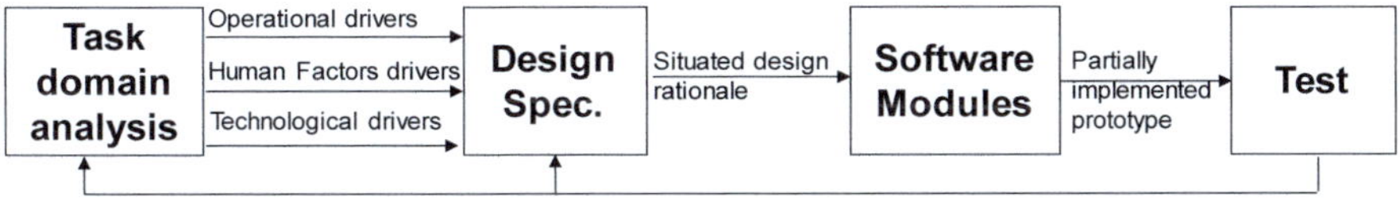

Fig. 1. Steps in the situated cognitive engineering methodology

In the first phase, we conducted a task domain analysis, and identified the most important operational, human, and technological drivers. From a technological perspective, we have identified predictive analytics as a crucial technology to enable a roaming operator. Predictive analytics can be used to predict future situations based on data from the past using machine learning algorithms. For example, to predict whether conditions are expected to remain stable, allowing the operator to leave, or to predict when alarms are likely to appear, requiring the operator to return to stationary position. From a human factors perspective, we identified a number of potential problems, related to trust, cognitive overload, and other issues well known in the HF literature [4]. For example, misalignments of operator's trust in the system could occur, because the performance of predictive analytics changes over time as more training data is used. These concerns must be adequately addressed in the design specification. In the second phase, we have specified the design specification which aims at providing a solution to the problems in the task domain. We will present these results in terms of user requirements, design patterns, and claims (which specify the rationale behind a design decision). The design specification is implemented in software modules in the third phase. We have implemented the most important patterns and user requirements in an early prototype of IOSS, which enables us to test if they bring about the expected results. To test IOSS (the fourth phase), we have set up a simulation environment, which allows us to give feedback on the earlier phases of the design process based on experiences in a semi-realistic end user environment.

The paper is structured around the four phases of the methodology. Section 2 describes the task domain analysis of the DP domain, followed by the design specification of IOSS in Sect. 3. Section 4 describes the software components and their interaction, followed by a description of the demonstrator which we used as a way of early testing IOSS (Sect. 4). The results of this test and our experiences while developing IOSS are described in the Sect. 5 (Conclusion).

2 Task Domain Analysis

As described in the introduction, the operational innovation we aim for is to allow an operator to be roaming and perform other tasks if conditions allow. This section describes relevant human factors and technological considerations when designing a system that allows these type of operations.

2.1 Human Factors

We have identified the following human factors issues to be relevant for IOSS.

Firstly, IOSS should address the issues of *maintaining operator's Situation Awareness* (SA). As the human operator is increasingly supported by intelligent technologies, the role of the human has evolved from direct system operator, to controller of automation, to supervisor of automation. Overall, this has had great positive effects on performance and costs, but research has shown that negative effects may arise due to lack of situation awareness [5] and out-of-the-loop problems [6]. This prevents operators from making effective decisions and causes errors [5]. These issues can lead to disastrous incidents in case automation fails. Particularly for roaming operators, we cannot take for granted that an operator's SA is at an appropriate level when (s)he returns to the task of operating DP after having been away for a while.

Secondly, IOSS must ensure the establishment of an *appropriate level of trust,* i.e. avoiding situations of overtrust and undertrust [7]. Overtrust occurs when the operator's trust in the system exceeds the system's capabilities. This situation could result in erroneous behavior as system mistakes are not corrected by the human supervisor. Undertrust occurs when operator trust falls short of the system's capabilities. This situation could lead to unnecessary operator workload, which in turn could lead to errors. As IOSS aims at establishing a higher degree of automation, the range of tasks that are performed by the system increases, and addressing trust concerns with respect to these tasks becomes even more important.

Thirdly, IOSS must establish an *appropriate level of cognitive task load (CTL),* avoiding both cognitive overload and underload. Cognitive overload occurs when the human cannot process all information that is provided by the system. An example of this in DP operations is known as alarm flooding, when the operator cannot timely respond to each alarm anymore [8], leading to suboptimal performance. Cognitive underload occurs, when the operator experiences insufficient workload, leading to drowsiness and inattentive behavior. This is a common problem for operators of highly automated DP systems, especially at night.

Fourthly, IOSS must adapt the interaction to the dynamic context of use, also known as *context aware interaction*. This issue becomes particularly relevant when we adopt the notion of a roaming operator. Because, the question of how often and in which way notifications should be sent highly depends on what the operator is doing [9].

2.2 Technological Drivers

We have identified a number of technological trends that will play a major role in future DP operations. Three of these technologies are outlined below.

Firstly, *predictive analytics techniques* are expected to have a major impact on the maritime world [10]. One possible application is predictive maintenance where large quantities of sensor data is collected and used as input for a machine learning algorithm. Over time, the algorithm should be able to recognize system failures before they occur, using data of the past. Such a classifier would be useful to our DP application where potential component failures are important to the operator. Many other applications of predictive analytics to DP are conceivable, for example, predicting position-loss based on weather data, or predicting operator's drowsiness based on physiological data (see [11] for an example in the automotive domain).

Secondly, *Internet of Things* [12] can be regarded as having a major impact in the maritime domain by allowing an unprecedented amount of data to be gathered and shared on a vessel. Virtually every component of a ship could become an information processing node in a large network. Applications in the DP domain could be monitoring the location of an operator, and disclosing vast amounts of additional information sources to the DP system to enable it to function more accurately.

Thirdly, computers are becoming more and more used as *personal assistants* (e.g., Siri[1], and google home[2]), which changes the relation between human and computer from that of a reactive tool to a more proactive entity (e.g. teammate [13]). As explained in the remainder of this paper, IOSS should be viewed as a personal assistant.

2.3 Combining Perspectives

The different drivers discussed in the previous sections are summarized in the following table (Table 1).

Table 1. Operational, human factors, and technological drivers

Operational	Human factors	Technological
Roaming operator	Maintaining SA	Predictive analytics
	Appropriate level of trust	Internet of Things (IoT)
	Appropriate level of CTL	Personal assistants
	Context aware interaction	

In this early phase of design, it already becomes apparent that the technological drivers of IOSS do not straightforwardly match with human factors drivers. For example, we could expect operator mistrust in a system that is based on predictive analytics algorithms. This is because the performance of such a system changes over

1 http://www.apple.com/ios/siri/.

2 https://madeby.google.com/home/.

time and is dependent on the amount of training data it has used. We cannot take for granted that the operator is capable of making proper judgements of the prediction's trustworthiness. Another problem could be information overload of the operator, caused by the massive amount of data made available by the IoT. Also, the use of mobile devices could lead to smaller graphical interfaces that convey fewer information than the stationary displays, having a direct effect on situation awareness.

A solution to these problems is proposed in the next Section.

3 Design Specification

Following the sCE methodology, the design specification is described from multiple perspectives. From a functional perspective, the Human-machine team functions are specified in terms of high level user requirements. From an interaction design perspective, different parts of the design solutions of IOSS are described in terms of design patterns. From an ontological perspective, the most important concepts and relations are defined that are used in the knowledge representation of IOSS [14].

3.1 Functional Perspective

Following the sCE methodology [15], the functional design is specified using use cases (that specify relevant environmental context, i.e. situatedness), core functions (specifying the main functionality of IOSS), and claims (specifying the reason why the function is required, i.e. design rationale).

An excerpt from envisioned core functions for IOSS is shown in Fig. 2:

Adaptive Automation
- IOSS should be adaptable w.r.t. task division and communication style
- IOSS should adapt its communication style according to user state
- IOSS should prevent cognitive overload of its user
- IOSS should behave according to a mixed initiative interaction style

User interface
- IOSS should support mobile and stationary UI's

Situational Awareness
- IOSS should support prediction of future situations
- IOSS should support change detection
- IOSS should support procedure awareness

Trust calibration
- IOSS should be able to explain itself
- IOSS should have a recognizable appearance

Agent architecture
- IOSS should be capable of acting in an open system
- IOSS should be capable of integrating information from multiple sources

Fig. 2. Excerpt of core functions of IOSS

The core functions are divided in five parts, each of which will be briefly discussed below. The requirements for *adaptive automation* aim to ensure a balanced workload which is tailored to the current situation of the user. This impacts the density of information that is communicated between user and IOSS, and finding a proper balance is regarded to be a responsibility of both, i.e. mixed initiative interaction. This means that the user is capable of instructing the computer when and how it wishes to be notified about which information by making *working agreements* [16]. The system also adapts its communication style to match the user's state (e.g., being brief when the operator is busy, and being more elaborate when the operator is not that busy). The *user interface* requirements state that both mobile and stationary user interfaces are needed to allow the concept of a roaming operator. The *Situation Awareness* requirements are intended to provide the operator with a sufficient level of SA [4]. At their most fine grained level (not shown in Fig. 2), these requirements specify exactly which information must be communicated in which types of situations. However, as stated above, these are adaptable to the user's preferences using working agreements. The requirements regarding *trust calibration* aim to prevent distrust by ensuring that the agent is capable of explaining the outcomes of the predictive analytics algorithms (i.e., explainable AI [18]). Because IOSS is used complementary to the DP-system (and its alarm system), a different trust relation should be built up with the DP system (which produces alarms that legally require a response [17]), and IOSS which learns over time and could mistakenly produce wrong predictions. To make it clear to the operator if he is interacting with IOSS or with the DP system, the IOSS must have a recognizable appearance. The last set of requirements deals with architectural issues, such as openness of the system, and access to digital information sources.

3.2 Interaction Design Perspective

Whereas the functional specification describes what IOSS should be capable of, the interaction design patterns specify how this must be established [19]. For IOSS, we have specified multiple design patterns. For example, one design pattern describes a notification (called a *smart notification*) in which interactive dialogues can take place to achieve explainability. Unlike an alarm, which contains a brief text statement about a problem, a smart notification presents the message in layers that can be exposed using a dialogue. Another design pattern describes how a user can deal with multiple smart notifications, set irrelevant notifications to inactive, and relevant notification to monitor mode. A detailed discussion of design patterns for human agent teams is beyond the scope of this paper. For more information about the specification and implementation of these design patterns, the reader is referred to [19]. In the following sections, the implementation in the software prototype is discussed in more detail.

3.3 Ontological Perspective

To realize the adaptable interaction between IOSS and the DPO, ontologies are required that specify the shared concepts that enable communication. IOSS utilizes several ontologies at different levels.

The most basic ontology behind IOSS can be seen in Fig. 3. This ontology defines several basic concepts, such as what an agent and action are and how these relate to each other. In addition, this ontology specifies the concept of a policy decision. A policy is an implementation of a working agreement in the form of *"if <condition> then <create PolicyDecision>"*. A *PolicyDecision* can be about prohibiting or obligating a certain *Action* for a certain *Actor* (e.g., that IOSS must initiate a dialogue with the DPO to communicate a certain piece of information.

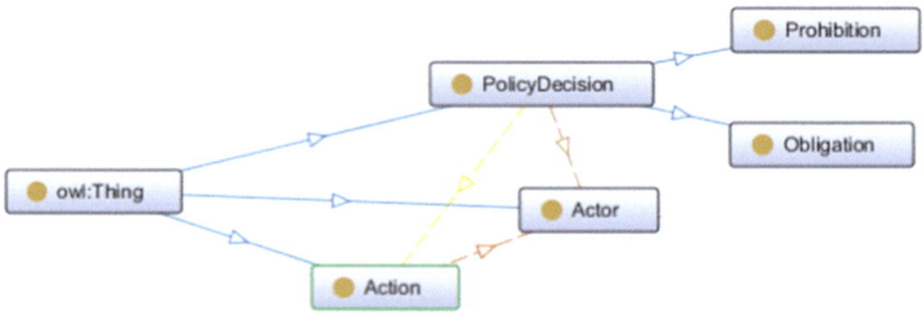

Fig. 3. One of the basic ontologies behind IOSS

Besides the basic ontology discussed above (which is very generic and abstract), domain specific ontologies are used to represent the knowledge required to create adequate working agreements. This removes the need for a large generic ontology of everything that is often difficult to comprehend. An example of a domain specific ontology is an ontology that defines the incoming sensor-data of the ship. With this, the user can create working agreements with IOSS that trigger on sensor values. Another ontology used by IOSS is the ontology of interactions (or *smart notifications*) that is used to describe the different ways in which the DPO can be informed.

4 Software Implementation

To implement the system described above, we have chosen to abandon the paradigm of alarm-based control systems for DP, in favor of the agent based paradigm [20]. An agent functions as a standalone component which monitors the user and provides assistance based on the current context. Instead of sending only factual information to its user, the agent has the capability of participating in a meaningful dialogue, express judgements, provide advice, or discuss remarkable situations. The agent may even be mistaken at times, just like humans. This is why it is important that the operator is allowed to develop accurate levels of trust in the agent. To facilitate this, the agent has a recognizable appearance, a so-called avatar.

For the IOSS, we designed an agent-based system architecture that allows multiple technologies to be combined. As shown in Fig. 4, we distinguish between three components within the social-technical system: the DP operator, the agent (IOSS) that supports the user, and the shared environment that contains digital information that can be accessed by both the operator and the agent.

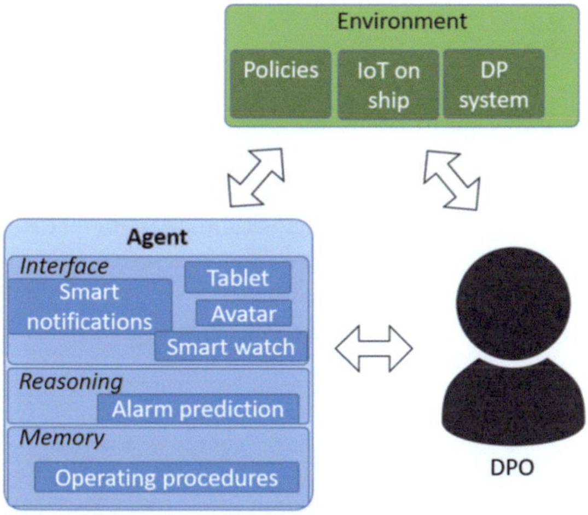

Fig. 4. Architectural overview of the human agent team.

The DP system remains unchanged and becomes part of the shared environment. In this way, all legacy systems are kept intact and IOSS is built on top as a layer of additional functionality. Other parts of the environment are the IoT (e.g., containing location tracking sensors) and a policy engine. As argued by Bunch et al. [21], a policy engine can be used to specify notification rules that allow users to adapt when and how the user is notified. We follow a similar approach and specify policies in the Drools expert system language[3]. An example of a policy in our case is:

`If wind speed is greater than 6 bft and the operator is roaming then IOSS must suggest to the operator to come back to stationary position`

An important feature of our policy engine is that these rules are understandable for non-programming experts, which allows them to adapt these rules to their liking.

4.1 Demonstrator

With the monitoring ability of IOSS, mobile devices, proximity sensors, and notifying devices we demonstrated that the concept of a roaming operator is feasible[4].

The agent, called IOSS, is the smart component that combines machine learning, IoT, and intelligent interfaces (as discussed in the previous section). We are currently developing the machine learning components with which it is possible to predict alarms in an early stage. That is to say, to learn to detect events, circumstances and weak signals which in the past have led to problems. With early detection on the basis of

[3] https://www.drools.org/.

[4] Human Enhancement by Maritime Adaptive Automation, TNO, https://youtu.be/MH0Vj-rChrM.

weak signals and data analytics, we hope to achieve that the window in which a problem is detected and solved can be enlarged and that the attention of the operator is aroused and is 'drawn' into the loop.

The IoT technology is used to locate the position of the operator on board of the vessel. For example, to determine whether the operator is sitting in front of the DP workstation or is roaming. If the DPO is roaming, what is the distance to the workstation, and what is the estimated 'return time' of the operator? This is important input for the intelligent interface which must decide if the interaction should take place on the operators' mobile device (a tablet or smartwatch), or on the stationary interface. Figure 5 depicts a screenshot of the interface where the agent (shown by the avatar in the lower left corner) engages in dialogue with its user.

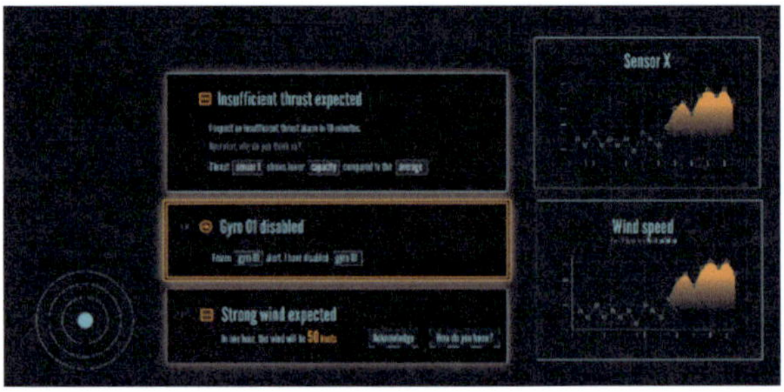

Fig. 5. Screenshot of IOSS

The next figure (Fig. 6) shows the use of IOSS in the stationary condition (the left screen), where the operator uses IOSS in combination with traditional DP interfaces (the two large screens on the right).

Fig. 6. IOSS in stationary condition

In stable conditions, the operator can be roaming and do other things. A photo of the roaming condition is provided below (Fig. 7).

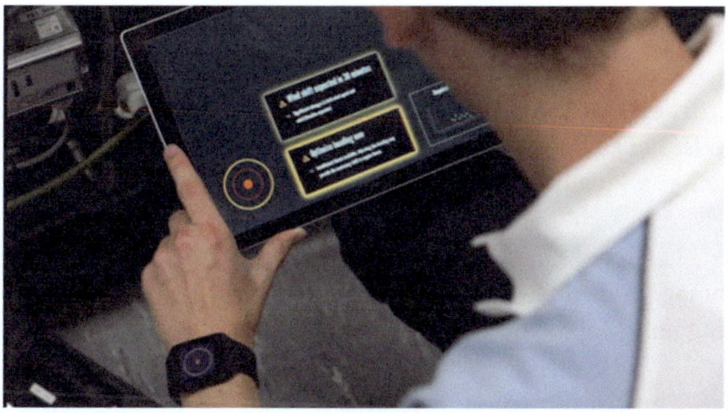

Fig. 7. Using IOSS in mobile condition

In the roaming condition, IOSS can advise the operator to return to stationary position, i.e. the bridge. If the operator is not looking at his screen (because he is busy doing other things), IOSS will notify the operator using a tactile signal on the smartwatch.

5 Conclusion

This paper proposes an intelligent operator support system (IOSS) for dynamic positioning systems. The IOSS functions at a high level of autonomy allowing the operator to be roaming in stable conditions. IOSS is aware of its limitations, and calls back the operator to stationary position in more demanding conditions. We have developed a first design specification and prototype and dealt with a number of (conflicting) operational, technological, and human factors demands. When designing a highly autonomous system such as IOSS, the classical risk of automation paradox (i.e. that the system's disadvantages overshadow its advantages) is very relevant. This paper proposes an agent-based approach that not only considers the problem-solving technology itself (e.g. *predictive analytics*), but also considers the technology that is required to team up with humans to avoid the automation paradox (e.g. *personal assistants* technology such as *smart notifications, working agreements; IoT* technology such as *location tracking, multiple mobile devices).* This paper demonstrates the possibility to combine these components in a meaningful way as a start to develop an IOSS that acts as a true teammate of DPO's.

Because the design of these different components is highly interdependent, much work remains to be done to evaluate and refine the working of IOSS. We are currently performing tests in a controlled end user environment allowing us to refine the system. Ultimately, these tests should prove that DP operations can be performed as safe with a roaming operator and IOSS as they can using a stationary operator.

References

1. Bainbridge, L.: Ironies of automation. Automatica **19**(6), 775–779 (1983)
2. Klein, G., Woods, D.D., Bradshaw, J.M., Hoffman, R.R., Feltovich, P.J.: Ten challenges for making automation a "team player" in joint human-agent activity. IEEE Intell. Syst. **19**(6), 91–95 (2004)
3. Neerincx, M.A., Lindenberg, J.: Situated cognitive engineering for complex task environments. In: Schraagen, J.M., Militello, L.G., Ormerod, T., Lipshitz, R. (eds.) Naturalistic Decision Making and Macrocognition, pp. 373–389. Ashgate Publishing Limited, Aldershot (2008)
4. Endsley, M.R.: From here to autonomy: lessons learned from human–automation research. Hum. Factors (2016). doi:10.1177/0018720816681350
5. Sarter, N.B., Woods, D.D.: How in the world did we ever get into that mode? Mode error and awareness in supervisory control. Hum. Factors **37**(1), 5–19 (1995)
6. Kaber, D.B., Endsley, M.R.: Out-of-the-loop performance problems and the use of intermediate levels of automation for improved control system functioning and safety. Process Saf. Prog. **16**(3), 126–131 (1997)
7. Lee, J.D., See, K.A.: Trust in automation: designing for appropriate reliance. Hum. Factors: J. Hum. Factors Ergon. Soc. **46**(1), 50–80 (2004)
8. Bullemer, P.T., Tolsma, M., Reising, D.V.C., Laberge, J.C.: Towards improving operator alarm flood responses: alternative alarm presentation techniques. Abnormal Situation Management Consortium (2011)
9. van Diggelen, J., Grootjen, M., Ubink, E.M., van Zomeren, M., Smets, N.J.: Content-based design and implementation of ambient intelligence applications. In: van Berlo, A., Hallenborg, K., Rodríguez, J., Tapia, D., Novais, P. (eds.) Ambient Intelligence-Software and Applications, pp. 1–8. Springer International Publishing, Heidelberg (2013)
10. Lee, H.G.: A study on predictive analytics application to ship machinery maintenance. Doctoral dissertation, Naval Postgraduate School, Monterey, California (2013)
11. Singh, H., Bhatia, J.S., Kaur, J.: Eye tracking based driver fatigue monitoring and warning system. In: 2010 India International Conference on Power Electronics (IICPE), pp. 1–6. IEEE, January 2011
12. Atzori, L., Iera, A., Morabito, G.: The internet of things: a survey. Comput. Netw. **54**(15), 2787–2805 (2010)
13. Bradshaw, J.M., Feltovich, P., Johnson, M., Breedy, M., Bunch, L., Eskridge, T., Jung, H., Lott, J., Uszok, A., Diggelen, J.: From tools to teammates: joint activity in human-agent-robot teams. In: Kurosu, M. (ed.) Human Centered Design, pp. 935–944. Springer, Heidelberg (2009)
14. Guarino, N.: Formal ontology, conceptual analysis and knowledge representation. Int. J. Hum. Comput. Stud. **43**(5–6), 625–640 (1995)
15. van Diggelen, J., van Drimmelen, K., Heuvelink, A., Kerbusch, P.J., Neerincx, M.A., van Trijp, S., van der Vecht, B.: Mutual empowerment in mobile soldier support. J. Battlefield Technol. **15**(1), 11 (2012)
16. Arciszewski, H.F., De Greef, T.E., Van Delft, J.H.: Adaptive automation in a naval combat management system. IEEE Trans. Syst. Man Cybern.-Part A: Syst. Hum. **39**(6), 1188–1199 (2009)
17. ANSI/ISA–18.2–2009: Management of Alarm Systems for the Process Industries (2009). http://ipi.ir/standard/STANDS/ISA/18.2.pdf

18. Van Lent, M., Fisher, W., Mancuso, M.: An explainable artificial intelligence system for small-unit tactical behavior. In: Proceedings of the National Conference on Artificial Intelligence, pp. 900–907. AAAI Press/MIT Press, Menlo Park/Cambridge/London, July 2004
19. Neerincx, M.A., Diggelen, J., Breda, L.: Interaction design patterns for adaptive human-agent-robot teamwork in high-risk domains. In: Harris, D. (ed.) Engineering Psychology and Cognitive Ergonomics, pp. 211–220. Springer, Cham (2016)
20. Wooldridge, M., Jennings, N.R.: Intelligent agents: theory and practice. knowl. Eng. Rev. **10** (02), 115–152 (1995)
21. Bunch, L., Breedy, M., Bradshaw, J.M., Carvalho, M., Danks, D., Suri, N.: Flexible automated monitoring and notification for complex processes using KARMEN. In: 2005 Proceedings of IEEE Networking, Sensing and Control, pp. 443–448. IEEE, March 2005

Sustainable Energy: Human Factors in Geothermal Water Resource Management

Barbara Tomaszewska(✉)

AGH University of Science and Technology,
Mickiewicza 30, 30-059 Krakow, Poland
barbara.tomaszewska@agh.edu.pl

Abstract. Investigating the knowledge, perceptions, as well as attitudes of the public concerning various aspects of environmental issues is of great importance in promoting sustainable development. An integrated understanding of the role of the effective use of renewable resources can provide proper support for the planning of the construction of new installations. In this sense, human factors and knowledge play an important role in conceptual models in the field of the effective use of renewable energy. This work is focused on the effective use of geothermal water and renewable energy for the future needs of modern sustainable, effective management of the geothermal water resources. The research field includes key new areas of study: an improvement in the management of geothermal and freshwater resources through the use of residual geothermal water and the development of the comprehensive utilization of water in agriculture, balneotherapy, and the tourist and leisure sectors.

Keywords: Sustainable energy · Human factors · Renewable resources · Geothermal water

1 Introduction

Damage to the ecosystems that support our existence is particularly troubling amidst the growing concern about climate change, loss of biodiversity, changes in worldwide biogeochemical cycles, land transformation, and potential decline in non-renewable energy resources on which we depend [1–3]. Tatcher and Yeow [3] showed that these interdisciplinary problems are complex and interconnected (Table 1). Also Dul et al. [4] noted that the human factor mainly takes a systems approach. We need to understand what is meant by a sustainable system. Wilson [5] commented that a system is an organised whole with interacting components contained within the boundaries of the whole. Looking at it from a different angle, Costanza and Patten [6] found a dictionary definition of a sustainable system which is quite straightforward, suggesting that it is simply a system that can persist indefinitely. However, this definition does not address the questions of which sustainable system needs to be considered and what time frame is considered sustainable from the point of view of the human factor. So, we can only believe that we are now at a point where we can start to demonstrate the practical applications of the human factor for a sustainable future.

P. Fechtelkotter and M. Legatt (eds.), *Advances in Human Factors in Energy: Oil, Gas, Nuclear and Electric Power Industries*, Advances in Intelligent Systems and Computing 599, DOI 10.1007/978-3-319-60204-2_7

Table 1. Ecological consequences of human activity (after [3]).

Ecological consequences	Human activity causes	Impacts
Climate change	Greenhouse gas accumulation (i.e. CO2, methane, etc.); aerosols	Change in rainfall patterns; desertification; excessive temperatures; severe weather systems; changes in disease distributions (e.g. malaria)
Loss of biodiversity	Alien invasion; monocultures through farming, agriculture, and industry; climate change	Species extinction; loss of potential medicines; loss of ecosystems; loss of resilience and adaptability of ecosystems; loss of bio-services
Biogeochemical disruptions	Disruptions to the carbon cycle; disruptions to the nitrogen cycle; freshwater pollution; drawing on ancient aquifers; increased concentrations of heavy metals, polychlorinated biphenyls (PCB's), and volatile organic compounds (VOCs)	Increased carbon in the atmosphere; acid rain; ocean acidification; freshwater pollution; respiratory problems; ozone depletion
Land transformation	Monocultures through farming, agriculture, and industry; urbanisation and built environments	Reduction in biomass through deforestation for agriculture, building materials and fuels; destruction of wetlands; overgrazing; ineffective nutrient circulation; freshwater scarcity; issues of sanitation; disruptions to the nitrogen cycle
Non-renewable energy resources	Peak and decline of energy resources; inefficient use of energy; inefficient transportation networks	Greenhouse gas accumulation; climate change; concentrations of nuclear waste
Population pressures	Over-population; land transformation	Social conflict; poor water quality; famine; lack of adequate sanitation; poverty; cultural intolerance

The best solution for reaching sustainable (harmonious) socio-economic development on the regional level is to diversify regional specialisation by using innovations and modern technologies based on endogenous resources. The efficient and sustainable management of natural resources, primarily energy resources, and increasing the share of renewable sources of energy are both goals of European Union Member States in the era of the fight against global warming. These objectives, especially economic ones, should undoubtedly be harmonised with environmental protection [7]. In this context, one important challenge is to explore and identify the opportunities for using renewable energy. Depending on local conditions, considerations are directed towards the use of solar energy, wind or geothermal energy. Each project always has an individual character. Investigating the knowledge, perceptions, as well as attitudes of the public concerning various aspects of environmental issues is of great importance to promoting sustainable development. An integrated understanding of the role of the effective use of renewable resources can properly support the planning of the construction of new installations. In this sense, human factors and knowledge play an important role in conceptual models in the field of effective usage of renewable energy.

Geothermal energy is used for power generation, space heating/cooling, buildings, soil and water (including water for aquaculture), the drying of crops and grain, and the heating of greenhouses. The use of geothermal energy, which is a constant heat source available 24 hour a day, 365 days a year, is often underestimated.

At the end of 2014, the total installed worldwide capacity from geothermal power plants reached 12,635 MWe, which corresponds to a worldwide production of 73,549 GWh of electricity per year [8]. The installed capacity on individual continents, as well as in individual countries, varies greatly (Fig. 1). A definite leader is the United States of America (USA). The present gross installed capacity for electricity is 3,450 MWe with 2,542 MWe net (running), producing approximately 16,600 GWh per year. Geothermal electric power plants are located in California, Nevada, Utah and Hawaii with recent installations in Alaska, Idaho, New Mexico, Oregon, and Wyoming [8].

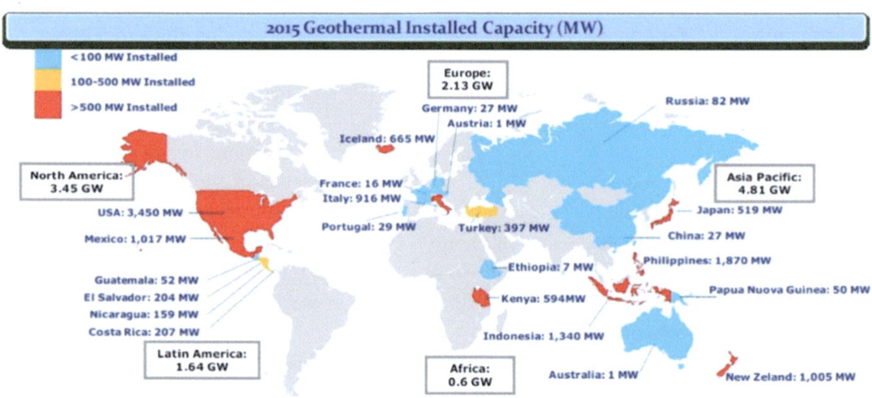

Fig. 1. Installed capacity in 2015 worldwide [12.6 GWe] (after [8]).

In addition to power generation, the main uses of geothermal waters in the world are space heating, district heating, spas, balneology, aquaculture and greenhouse heating [9]. Direct utilisation of geothermal energy in a total of 82 countries is an increase from the 78 reported in 2010, 72 reported in 2005, 58 reported in 2000, and 28 reported in 1995. An estimate of the installed thermal power for direct utilisation of geothermal energy at the end of 2014 is 70,329 MWt, almost a 45% increase over the 2010 data, growing at a compound rate of 7.7% annually with a capacity factor of 0.265. The thermal energy used is 587,786 TJ/yr (163,287 GWh/yr), about a 38.7% increase over 2010, growing at a compound rate of 6.8% annually [9].

There are a lot of examples of this type of system around the world and of different technologies used [9–11], depending on the temperature of the geothermal resources. Also some methods of analysis of the main advantages and disadvantages of the systems are presented in the literature. But, in many cases, apart from the technological, and also legal and financial elements, the human factor plays a crucial role in the field of promotion, multivariate optimisation, and comprehensive utilisation of the resource potential [12].

This work is focused on the effective use of geothermal water and renewable energy for future needs in modern sustainable and effective management of geothermal water resources. The main objective of the proposed solution is to determine the potential for the cascade use of geothermal resources, and also the direct coupling of the use of geothermal energy with desalination processes. The key issue in the research is the presentation of the human factors in planning sustainable energy and water management. The research field includes key new areas of study: an improvement in the management of geothermal and freshwater resources through the use of residual geothermal water and the development of the comprehensive utilisation of water in agriculture, balneotherapy, and the tourist and leisure sectors.

2 Cascade Systems for the Direct Utilisation of Geothermal Heat

Sustainable and efficient management of geothermal energy should be focused on the comprehensive utilisation of resources obtained and extracted from the reservoir. Designs promoting energy efficiency should be targeted to achieve the highest possible Δt (t_1–t_2), the difference between the outflow water temperature (t_1) and temperature of the cooled geothermal water (t_2). In many cases a cascade system for using geothermal energy is the best solution for its effective utilisation. Space heating requires temperatures in the range 50–100 °C, with 40 °C useful in some marginal cases and ground-source heat pumps extending the range down to 5 °C. Cooling and industrial processing normally require temperatures over 100 °C [11, 13]. Agricultural and aquacultural uses require the lowest temperatures, with values from 25 to 90 °C.

Agricultural and agro-industrial uses form a very important part of geothermal energy applications [14]. Popovski [15] identified four types of direct application of geothermal energy in agriculture: (1) greenhouse heating; (2) aquaculture (fish farming and algae production); (3) agro-industrial processes; (4) soil heating (of open-field plant root systems). Geothermal waters were first used in greenhouse heating in Iceland in the 1920s [16]. In Greece, the use of geothermal waters for greenhouse heating started in the early 1980s [17] and in the cultivation of Spirulina in the late 1990s. In the region of Klamath Falls (Oregon, USA) a system with a temperature of 27 °C to 93 °C makes use of a cascade for agriculture and aquaculture. The main well is used for greenhouse heating at 93 °C. The use of geothermal energy for heating and agriculture in a cascade has also been carried out in the Mineral and Energy Economy Research Institute of the Polish Academy of Sciences (PAS MEERI) in Bańska Niżna [18]. The first geothermal plant in Poland was constructed and put into operation in the Podhale Basin in southern Poland. The cascaded heat supply included five stages of heat distribution based upon a secondary circulation loop. The first step provides hot water for district heating for apartments and buildings and hot water for domestic use (85–65 °C). After the first thermal utilisation, the water had a temperature of 60–65 °C, which was enough for a timber drying building. The next steps included a parapet greenhouse (45 °C), a fish farm for thermophilus fish (35 °C) and the final step provided ground heating (30 °C) for polytunnels [18] (Fig. 2). This arrangement permits studies to be carried out on the multidirectional development of geothermal energy.

Fig. 2. The cascade system for the utilisation of geothermal energy in PAS MEERI in Poland.

The use of geothermal energy in cascade systems, however, requires the integration of activities and the involvement of interdisciplinary teams of people. In industrial-scale solutions, each element of the cascade is a separate element of economic activity that, in certain circumstances, requires implementation in close proximity. So, the human factor is the issue of the integration and co-operation of designers in the fields of geology, hydrogeology, energy engineers and potential businessmen - investors interested in the recovery of geothermal energy in new recreational facilities, agriculture, aquaculture and other fields. Local environmental values, as well as the supply and demand for manufactured products, play an important role. Geothermal energy management in cascading systems is therefore a complex system of environmental and economic interactions. Sustainable energy use requires sustainable investment planning, potential business partnerships, and cooperation to enable all stakeholders to achieve their goals. According to Szymańska and Chodkowska-Miszczuk [19] the development of renewable energy is based on the dissemination of innovations.

It is thus assumed that the initiators of renewable energy development are of working age. Human capital (age, education) contains stimuli for innovation. The young, better educated and mobile are often open-minded and creative and become leaders who introduce innovations [19].

3 Cascade Systems for the Direct Utilisation of Geothermal Water

Electricity production with heat recovery for space heating is well known where there are high enthalpy sources of geothermal energy, with examples from Iceland, USA, Italy, and other countries. But following the initial use of the geothermal steam and water, its secondary use is, in effect, making use of resources which should be used in a balanced manner. The most popular spa in Iceland, the Blue Lagoon, uses the same geothermal water that is exploited for the Svartsengi Power Station (Fig. 3). The electric power station, next to Reykjavik, was the world's first geothermal power plant for electric power generation and hot water production for district heating [20]. Apart from the direct method of electricity production, a binary system based on the Organic Rankine Cicle (ORC) is here exploited to promote the use of clean energy. The total generating capacity of the plant is 46.4 MW_e and 200 MW_{th} (2700TJ/year) in the form of hot water for the district heating [21]. Water rich in minerals like silica and sulphur is used in one of Iceland's popular bathing resorts, the Blue Lagoon (landscape: a permeable lava field) which is the most visited attraction in the country.

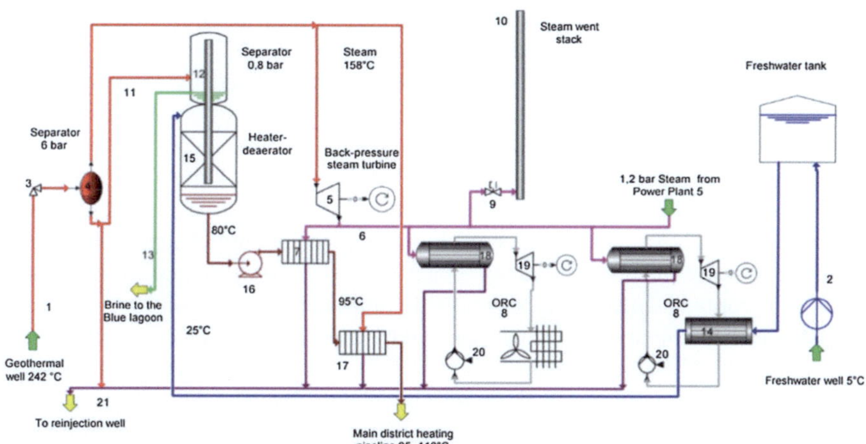

Fig. 3. Svartsengi power plant flow diagram (after [21]).

For low-enthalpy geothermal resources, the warm water can first be run through a space heating installation and then cascaded to swimming pools, greenhouses and/or for heating aquaculture ponds. This type of project maximises the use of the resource as well as improving the economics, geothermal resource management and sustainability. Key factors that determine the conditions in which geothermal waters are used, the amount of energy obtained and the manner in which cooled water is utilised include the water salinity and the presence of specific minerals. The geothermal waters present in the geological structures are fresh waters and waters with low mineral content as well as brines with mineral contents exceeding 100 g/L [22, 23]. In consequence, the salinity of geothermal waters varies widely. Elevated salinity levels and the presence of microelements such as boron, barium, strontium, fluorides, bromides and heavy metals [24] may often lead to difficulties related to the utilization of spent waters. Low mineral content and fresh geothermal waters are mostly made available for heating and leisure purposes. So, depending on the mineralisation of geothermal waters, the cascade system can also include balneotherapeutic and leisure purposes, not only in the field of energy use for heating pools but especially in relation to water utilisation. One should note that people have been using geothermal baths to stay healthy and feel good for thousands of years. This is why balneology, i.e. using natural spring geothermal water for medicinal purposes, has a very long history [25]. Ancient Greeks, Turks and Romans were known for developing bathing facilities practically everywhere where they went with their conquests. Baths were taken from Persia to present-day England [25]. The 21st century has brought a new approach to "travelling for health" and to treatment itself. Economic development and the related rise in the standard of living have changed people's attitude to their appearance, fitness and state of mind [25]. Nowadays, classic health resorts using balneological therapy have to satisfy the expectations of more and more demanding customers. The geothermal health resort is classically located in places where medical mineral water, especially natural hot water, medical peloids and natural gases, are used for bathing, drinking, inhalations, water jets, jacuzzi, body masks, etc. Balneotherapy, hydrotherapy and climatotherapy are the main (and basic) elements of healing in these resorts [25, 26].

The therapeutic value of geothermal waters is determined by their temperature, variety of dissolved ions, gases and trace elements. In Poland, groundwater which is not contaminated and containing natural variations in physicochemical parameters and at least one specific pharmaco-dynamic factor such as 2 mg F^-/L, 1 mg I^-/L, 1 mg S(II)/L, 70 mg H_2SiO_3/L, 10 mg Fe(II)/L, 74 Bq of radon/L, 250–1000 mg free CO_2/L and/or has a temperature above 20 °C is then considered as therapeutic water [27]. Moreover, whole mineral water (mineralization more than 1 g/L) can be medicinal water, regardless of whether it contains specifically mentioned components. Based on temperature, balneologists classify therapeutic waters as: cold (<20 °C), hypothermal (20 < 30 °C), thermal (30 < 40 °C) and hyperthermal (>40 °C). The use of geothermal waters for medicinal purposes in European countries is enhanced by health resorts employing qualified medical staff cooperating with hospitals, university hospitals and research institutes. This certainly raises the prestige of health resorts. At the same time, limiting the financial resources allocated for health resort treatment (provided by the national health care systems of specific countries) reduces interest in such treatment [25].

4 Geothermal Waters as a Source of Freshwater

Geothermal systems not only provide a valuable renewable energy, but can also be considered as the source and solution for fresh water production. Water is the basis of life and culture. Nowadays, when a common problem in the world is a shortage of drinking water, research plays a very important role in the better utilisation of geothermal water and its wastes. First of all, geothermal water can be an energy source but after treatment processes can then be used as a source of drinking water and/or for the public water supply. Crucially, the implementation of such solutions on an industrial scale would reduce the negative impact of the discharge of saline waste geothermal water to streams. Waste recycled in this way could be reused, which is very important, especially in areas with problems of a lack of water or water deficit. The most important application of research is the recycling of waste geothermal water and profiting from them due to the specific micro- and macroelement content, good quality, and lack of salinity components. Waters with a mineralisation below 1,000 mg/L normally meet the requirements for drinking, potable, and sanitary water. High quality geothermal waters are exploited as an energy and fresh water source in Mszczonów (central Poland). Water with a low mineral content (ca. 0.5 g/L) and with an intake temperature of 42 °C has been extracted since 2000 from the Mszczonów IG-1 well (Lower Cretaceous horizon composed of sandstones interbedded with mudstone and claystone). These are high-quality Cl-HCO_3-Na-Ca waters that are fed into the municipal water supply network as drinking water following cooling and simple treatment [22]. A similar method of utilising of geothermal waters is envisaged in Poddębice (central Poland) where geothermal water exhibits a mineralization of ca. 0.4 g/L at the intake. There are also some examples of using geothermal water as a drinking source in other European countries. Borovic and Markovic [28] present an analysis of the total geothermal water resource in Croatia, in which five natural springs and one deep borehole are directly exploited as sanitary water (e.g. in Stubičke toplice), one natural spring and one deep borehole are used for the public water supply and one borehole is used for table water. Geothermal water is also bottled as table and mineral water.

Gude [11] suggests that in many cases the cascade system of geothermal energy use can include desalination plants. Figure 4 presents an example of a design for an integrated configuration to produce power, followed by desalination using both thermal and membrane processes, then for applications in food processing, refrigeration plants and district heating or cooling systems, space heating of buildings, greenhouses and soil heating, industrial process heat, agricultural drying, and fish farming. One of the first proposals for the treatment of geothermal water for the purpose of producing potable water was presented by Houcine et al. [29]. The brackish geothermal groundwater (2.8 g/L, 30 °C) of the Chott El-Fejij (70 km from the city of Gabes, Tunisia) was used to feed the reverse osmosis desalination plant in the city of Gabes at a flow rate of 2,000 m^3/h. The authors explain the advantage of the productivity of RO membranes (water flux) with increasing temperature of geothermal feed water, as long as the temperature tolerance of the membrane is respected. The viscosity of water decreases with increasing water temperature and in consequence membrane productivity increases about 2–3% per 1 °C. This is an important fact because really small water resources occur in African countries where more than 70% of the population does not have access to potable water [30]. The results of a study using an electrical resistivity model aimed at investigating the potential of geothermal and groundwater aquifers in the Mayo Kani area (Cameroon) to contribute to the improvement of the living conditions of the population have been presented by Kana et al. [30]. The presence of two potential low enthalpy geothermal reservoirs, which can also be used as a potable water source, have been discovered at Djangal and Moundjou. In Eburru (Kenya), the cascade utilisation of geothermal water proposed includes the much

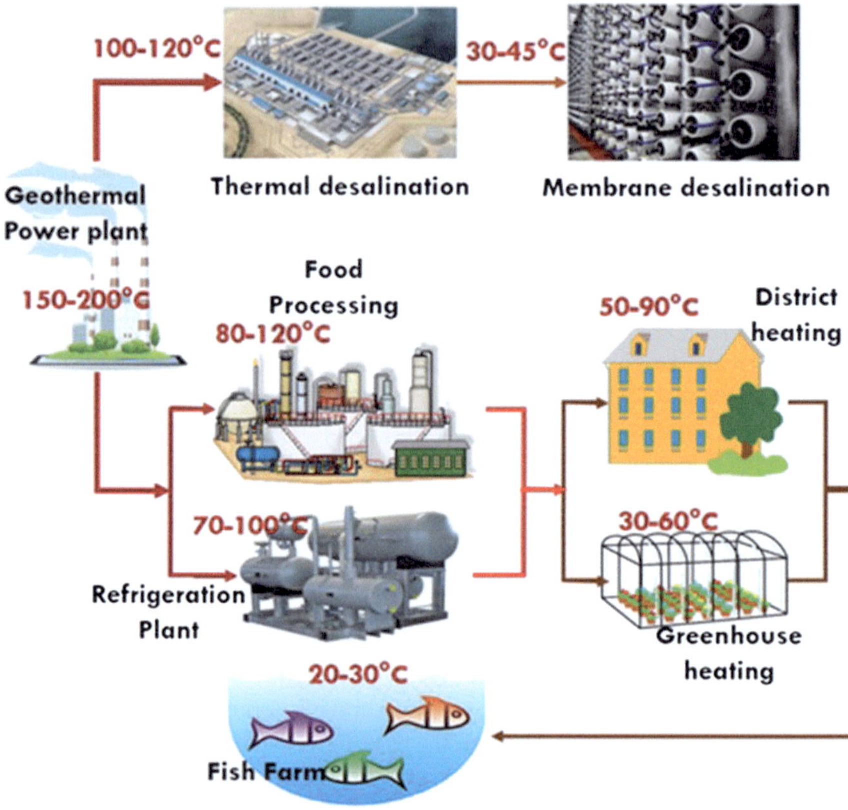

Fig. 4. Integrated configurations for geothermal energy sources – polygeneration for multiple benefits (after [11]).

needed potable water. Mburu [31] suggests that this will be used as a demonstration centre for the utilisation of geothermal energy as well as a source of income to the local community. Countries which have good geothermal conditions are ideal candidates for producing fresh water from brackish water [32].

Given the increasing deficit of freshwater worldwide, opportunities should be considered for desalinating and treating geothermal waters for drinking and household purposes. Nowadays, in countries with warm climates, cooled and treated geothermal water is mainly used for the irrigation of agricultural crops [33, 34]. Pilot studies have been carried out in Poland relating to the evaluation of the possibilities of using selected membrane techniques for desalinating geothermal waters [35]. The results of investigations have demonstrated that, after treatment, spent geothermal water may provide an alternative resource leading to the decentralisation of the supply of water for drinking and household purposes. It has been demonstrated that the use of a hybrid system based on iron removal, ultrafiltration, and reverse osmosis with BWRO membranes enables water (permeate) of high quality to be obtained. Taking into account the low pressure of 1.1 MPa used at the reverse osmosis stage, an efficient and stable desalination process was achieved for geothermal waters with a mineralisation of up to 7 g/L. with relatively high retention factors [35, 36]. The results of the quality tests conducted for the treated geothermal waters with respect to the requirements for water intended for human consumption have demonstrated the compliance with these requirements in terms of physical and chemical, microbiological and radiological indicators. In this context, the extraction of geothermal energy and the efficient and rational management of spent water may result in a number of benefits – not only in terms of energy and economy, but of social impact as well [22]. A very important aspect when undertaking such measures is the sustainable management of natural resources and respecting all elements of the natural environment at every stage of business activity.

5 Conclusions

In order to meet future energy policy targets, effective energy systems based on geothermal water need to use more environmentally friendly management. However, the successful and widespread implementation of the proposed use of geothermal energy in a cascade has been presented as a new approach - geothermal water as an important source for fresh water production. The examples presented provide guidelines for the engineering community and stakeholders at decision-level regarding an acceptable design from a user's perspective and the relevance of the technical characteristics such as the security of the energy source and at the same time the heat source for spa use, balneology, aquaculture and greenhouse heating, and fresh water supply.

It must be emphasised that the most crucial factor in determining sustainable geothermal water management is the human capital. In this sense, the human factor is the issue of the integration and co-operation of designers in the fields of geology, hydrogeology, energy engineers and potential businessmen - investors interested in the recovery of geothermal energy in new recreational facilities, agriculture, aquaculture and other fields. Geothermal energy and the management of geothermal water in cascade systems is therefore a complex system of environmental and economic interactions. Sustainable energy use requires sustainable investment planning, potential business partnerships, and cooperation to enable all the stakeholders to achieve their goals.

Acknowledgments. The paper has been prepared under the AGH-UST statutory research grant No. 11.11.140.321.

References

1. Vitousek, P.M., Mooney, H.A., Lubchenco, J., Melillo, J.M.: Human domination of Earth's ecosystems. Science **277**, 494–499 (1997)
2. Beddoe, R., Costanza, R., Farley, J., Garza, E., Kent, J., Kubiszewski, I., Martinez, L., McCowen, T., Murphy, K., Myers, N., Ogden, Z., Stapleton, K., Woodward, J.: Overcoming systemic roadblocks to sustainability: the evolutionary redesign of worldviews, institutions, and technologies. Proc. Natl. Acad. Sci. **106**, 2483–2489 (2009)

3. Tatcher, A., Yeow, P.H.P.: Human factors for a sustainable future. Appl Ergon. **57**, 1–7 (2016)
4. Dul, J., Bruder, R., Buckle, P., Carayon, P., Falzon, P., Marras, W.S., Wilson, J.R., van der Doelen, B.: A strategy for human factors/ergonomics: developing the discipline and profession. Ergonomics **55**, 377–395 (2012)
5. Wilson, J.R.: Fundamentals of systems ergonomics/human factors. Appl. Ergon. **45**, 5–13 (2014)
6. Costanza, R., Patten, B.C.: Defining and predicting sustainability. Ecol. Econ. **15**, 193–196 (1995)
7. Tomaszewska, B., Rajca, M., Kmiecik, E., Bodzek, M., Bujakowski, W., Wator, K., Tyszer, M.: The influence of selected factors on the effectiveness of pre-treatment of geothermal water during the nanofiltration process. Desalination **406**, 74–82 (2017)
8. Bertani, R.: Geothermal power generation in the world 2010–2014 update report. In: Proceedings World Geothermal Congress 2015, 19–25 April 2015, Melbourne, Australia. Paper 1001 (2015)
9. Lund, J.W., Boyd, T.L.: Direct utilization of geothermal energy 2015: worldwide review. In: Proceedings World Geothermal Congress 2015, 19–25 April 2015, Melbourne, Australia. Paper 1000 (2015)
10. Rubio-Maya, C., Díaz, V.A., Martínez, E.P., Belman-Flores, J.M.: Cascade utilization of low and medium enthalpy geothermal resources − a review. Renew. Sust. Energ. Rev. **52**, 689–716 (2015)
11. Gude, V.G.: Geothermal source potential for water desalination – current status and future perspective. Renew. Sust. Energ. Rev. **57**, 1038–1065 (2016)
12. Fechtelkotter, P., Bleakley, G., Douglass, B.P., Amaba, B.: Model-driven development for safety-critical projects in intelligent energy. In: SPE Middle East Intelligent Energy Conference and Exhibiiton, 28–30 October 2013, Dubai, UAE, SPE SPE-167479-MS, pp. 1–14 (2013)
13. Lund, J.W.: Direct utilization of geothermal energy. Energies **3**(8), 1443–1471 (2010)
14. Van Nguyen, M., Arason, S., Gissurarson, M., Pálsson, P.G.: Uses of Geothermal Energy in Food and Agriculture – Opportunities for Developing Countries. FAO, Rome (2015)
15. Popovski, K.: Agricultural and industrial uses of geothermal energy in Europe. In: Proceedings of the International Geothermal Days Slovakia 2009 – Conference and Summer School, 26–29 May 2009, Castá-Papiernicka, Slovakia (2009)
16. Barbier, E.: Geothermal energy technology and current status: an overview. Renew. Sust. Energ. Rev. **6**(1–2), 3–65 (2002)
17. Andritsos, N., Dalabakis, P., Karydakis, G., Kolios, N., Fytikas, M.: Characteristics of low-enthalpy geothermal applications in Greece. Renew. Energ. **36**, 1298–1305 (2011)
18. Bujakowski, W.: The pilot project of a geothermal heat recuperation cascade system for fish and vegetable breeding, the PAS MEERI geothermal laboratory (Podhale region, Poland). In: Proceedings of the World Geothermal Congress, 2000, 28 May–10 June 2000, Kyushu-Tohoku, Japan, pp. 3379–3383 (2000)
19. Szymanska, D., Chodkowska-Miszczuk, J.: Endogenous resources utilization of rural areas in shaping sustainable development in Poland. Renew. Sust. Energ. Rev. **15**, 1497–1501 (2011)
20. Lund, J.W., Chiasson, A.: Examples of combined heat and power plants using geothermal energy. In: Proceedings of the European Geothermal Congress, 30 May–1 June 2007, Unterchaching, Germany (2007)
21. Thorolfsson, G.: Sudurnes regional heating Corporation Svartsengi. Iceland. Geo-Heat Center Q. Bull. **26**(2), 14–18 (2005)

22. Tomaszewska, B., Szczepański, A.: Possibilities for the efficient utilisation of spent geothermal waters. Environ. Sci. Pollut. R. **21**, 11409–11417 (2014)
23. Tomaszewska, B., Bodzek, M.: The removal of radionucl, ides during desalination of geothermal waters containing boron using the BWRO system. Desalination **309**, 284–290 (2013)
24. Kmiecik, E., Tomaszewska, B., Wątor, K., Bodzek, M.: Selected problems with boron determination in water treatment processes. Part I: comparison of the reference methods for ICP-MS and ICP-OES determinations. Environ. Sci. Pollut. R. **23**(12), 11658–11667 (2016)
25. Kiełczawa, B.: Zarys balneoterapeutycznego zastosowania wód geotermalnych [Use of geothermal springs in balneology]. Wydz. Geoinżynierii, Górnictwa i Geologii Politechniki Wrocławskiej, Wrocław, Poland (2016)
26. Gutenbrunner, C., Bender, T., Cantista, P., Karagülle, Z.: A proposal for a worldwide definition of health resort medicine, balneology, medical hydrology and climatology. Int. J. Biometeorol. **54**, 495–507 (2010)
27. Wątor, K., Kmiecik, E., Tomaszewska, B.: Assessing medicinal qualities of groundwater from Busko-Zdrój area (Poland) using the probabilistic method. Environ. Earth Sci. **75**(804), 1–13 (2016)
28. Borovic, S., Markovic, I.: Utilization and tourism valorisation of geothermal waters in Croatia. Renew. Sust. Energ. Rev. **44**, 52–63 (2015)
29. Houcine, I., Benjemaa, F., Chahbani, M.H., Maalej, M.: Renewable energy sources for water desalting in Tunisia. Desalination **125**, 123–132 (1999)
30. Kana, J.D., Djongyang, N., Raidandi, D., Nouck, P.N., Nouayou, R., Tabod, T.C., Sanda, O.: Geophisical investigation of low enthalpy potential and ground water reservoirs in the Sudano-Sahelian region of Cameroon. J. Afr. Earth Sci. **110**, 81–91 (2015)
31. Mburu, M.: Cascaded use of geothermal energy: eburru case study. Geo-Heat Center Q. Bull. **30**(4), 21–26 (2012)
32. Mahmoudi, H., Spahis, N., Goosen, M.F., Ghaffour, N., Drouiche, N., Ouagued, A.: Application of geothermal energy for heating and fresh water production in a brackish water greenhouse desalination unit: a case study from Algeria. Renew. Sust. Energ. Rev. **14**(1), 512–517 (2010)
33. Koseoglu, H., Harman, B.I., Yigit, N.O., Guler, E., Kabay, N., Kitis, M.: The effects of operating conditions on boron removal from geothermal waters by membrane processes. Desalination **258**, 72–78 (2010)
34. Kabay, N., Yilmaz-Ipek, I., Soroko, I., Makowski, M., Kirmizisakal, O., Yag, S., Bryjak, M., Yuksel, M.: Removal of boron from Balcova geothermal water by ion exchange-microfiltration hybrid process. Desalination **241**, 167–173 (2009)
35. Tomaszewska, B., Pająk, L., Bodzek, M.: Application of a hybrid UF-RO process to geothermal water desalination. Concentrate disposal and costs analysis. Arch. Environ. Prot. **40**(3), 137–151 (2014)
36. Tomaszewska, B., Pająk, L.: Geothermal water resources management - economic aspects of their treatment. Miner. Resour. Manag. **4**, 59–70 (2012)

Autonomous Algorithm for Safety Systems of the Nuclear Power Plant by Using the Deep Learning

Daeil Lee and Jonghyun Kim(✉)

Department of Nuclear Engineering, Chosun University, 309 pilmun-daero, Dong-gu, Gwangju 501-709, Republic of Korea
dleodlf1004@Chosun.kr, jonghyun.kim@Chosun.ac.kr

Abstract. This study aims to develop an autonomous algorithm to control the safety systems of nuclear power plant (NPP) by using the deep learning that is one of machine learning methods. The autonomous algorithm has two main goals. First, it achieves a high level of automation for nine safety functions of NPP. Second, the algorithm controls the nine safety functions in an integrated way. The function-based hierarchical framework is suggested to represent the multi-level structure that models NPP safety systems with the levels of goal, function and system. The function-based hierarchical framework is used to model the NPP for the application of the multi-system deep learning network. Multi-system deep learning network is applied to develop the algorithm for autonomous control. This approach enables the systematic analysis of power plant system and development of the database for the deep learning network.

Keywords: Systems engineering · Deep-learning · Autonomous algorithm

1 Introduction

With the help of artificial intelligence, the autonomous control is being applied to many industries, e.g., power system [1], autonomous ground vehicles (AGV) [2, 3], autonomous underwater vehicles [4] and robotics [5]. The AlphaGo and the google car are the representative examples. The AlphaGo has a go ability equivalent to a professional human Go player through the deep learning algorithm [6]. The google car is aiming at the autonomous operation system without a driver.

Control systems with high degrees of autonomy have power and ability for self-governance in the performance of control functions. To achieve significantly higher degrees of autonomy, the controller must be able to perform a number of functions as well as the conventional control functions such as tracking and regulation [7].

Nuclear power plants (NPPs) are an industrial area where automatic (but not autonomous) control plays an important role in the safety and operation. The automation is applied due to many different reasons such as regulatory requirements, improvement of efficient, reduction of human errors, and harsh environmental condition. Especially, safety functions are those that are performed by the automatic control system. Goals of safety functions are shutting down the reactor, removing the heat and

P. Fechtelkotter and M. Legatt (eds.), *Advances in Human Factors in Energy: Oil, Gas, Nuclear and Electric Power Industries*, Advances in Intelligent Systems and Computing 599, DOI 10.1007/978-3-319-60204-2_8

preventing radiation release. There are some reasons for the automation of safety functions in NPPs as follows:

- Regulation requires automatic actuation of safety functions.
- Those functions require high reliability, accuracy, and speed for which the automatic system is known to be better than human operators.

This application of automation is even more highlighted in a current trend of nuclear industry, which is the development of small modular reactors (SMRs). An SMR has advantages of safety, convenient construction and plant operation compared to large nuclear power plants [8]. SMRs are to be constructed in the area where peoples are difficult to access (e.g. space, island, desert, polar regions). Therefore, SMRs require the high level of automation that is the self-operation or autonomous operation. Some studies have already been conducted to increase the automation level of NPPs to the autonomous control [9, 10].

However, the level of automation for the safety functions of current NPPs does not reach the autonomous control, because the intervention by operators is important in a large portion of control. For instance, although the actuation of systems is automated, maneuvering, reset, and termination still need to be performed by operators. In addition, the functions are operating independently, not interrelated functionally, even if they have a common goal, i.e., the safety of NPPs.

This study aims to develop an autonomous algorithm to control the safety systems of NPP by using the deep learning network that is one of the machine learning method. The autonomous algorithm has two main goals. First, it achieves a high level of automation for nine safety functions of NPP. Individual safety functions can be operated without any intervention of operators unless the autonomous control is abnormal in itself. Second, the algorithm controls the nine safety functions in an integrated way. It monitors/initiates/adjusts/terminates the functions interactively to achieve the goal of safety. This study suggests a conceptual design for the autonomous control in achieving the safety in NPPs. Section 2 shows the hierarchical framework and process used for analysis to complex systems of NPPs. In addition, it introduces a structure of the multi-system deep learning network. Section 3 presents the multi-system deep learning network and the hierarchical framework for the safety system in a NPP.

2 Approach

Function-based hierarchical framework for modeling systems and concept of multi-system deep learning network have been applied to autonomous algorithm design of NPP safety systems. The hierarchical framework was suggested to analyze complex systems of a NPP. The framework is used to model the NPP for the application of the deep learning network. Multi-system deep learning network is applied to develop the algorithm for autonomously controlling non-linear parameters of a NPP.

2.1 Function-Based Hierarchical Framework

This study suggested a function-based hierarchical framework to analyze a complex structure of NPP safety systems. The framework is the multi-level structure that models NPP safety systems with the hierarchical levels of goal, function and system, as shown in Fig. 1.

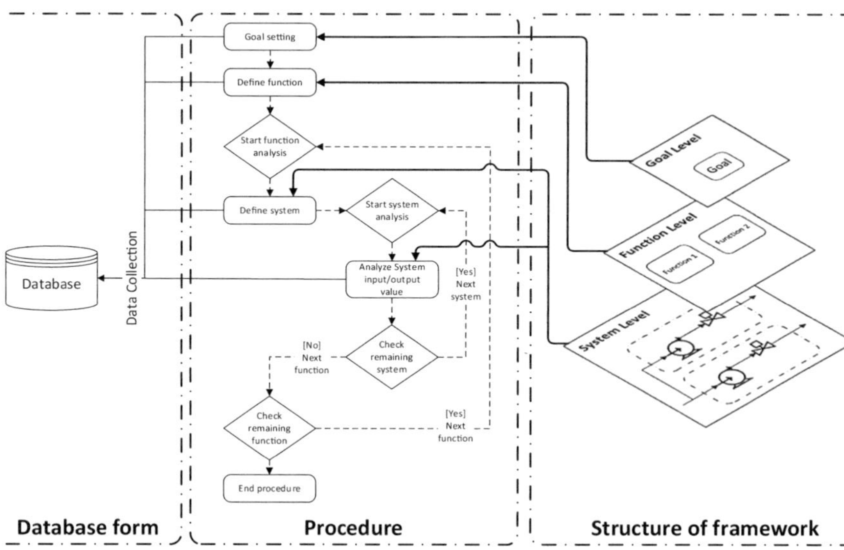

Fig. 1. Function-based hierarchical framework

The function-based hierarchical framework consists of three parts: structure, procedure and database form. The structure of function-based hierarchical framework aims at representing the hierarchical relation of goal-function-system for achieving the safety of NPPs. Goal level means the ultimate goal, i.e., safety of NPP. Function level represents the functions that should be satisfied to reach the goal. The function includes reactivity control, heat removal, and pressure/temperature control in NPPs. System level represents systems that the NPP implements to satisfy functions. The structure of framework can provide an understanding of the overall system.

The procedure is a sequential process for transforming the information of structure on the safety goals, functions, and systems to the database. The process consists of nine steps. The description of each step is presented in Table 1.

The database includes the information on the structure of safety goals, functions, and systems as shown in Fig. 2. This database is used for modeling the NPP in applying the deep learning network, providing the input/output of each system network in multi-system network. The database includes goal, function, system, function redundancy of system, component of system and required values (input, output) of system. The analyzed information defines goal, function, and system corresponding to each level of structure and considers the functional redundancy of the system. Functional redundancy means that a system can be used for more than one function.

Table 1. Procedure of nine steps

Step number	Content
(1)	Goal setting
(2)	Define function
(3)	Start function analysis
(4)	Define system
(5)	Start system analysis
(6)	Analyze system input/output value
(7)	Check remaining system
(8)	Check remaining function
(9)	End procedure

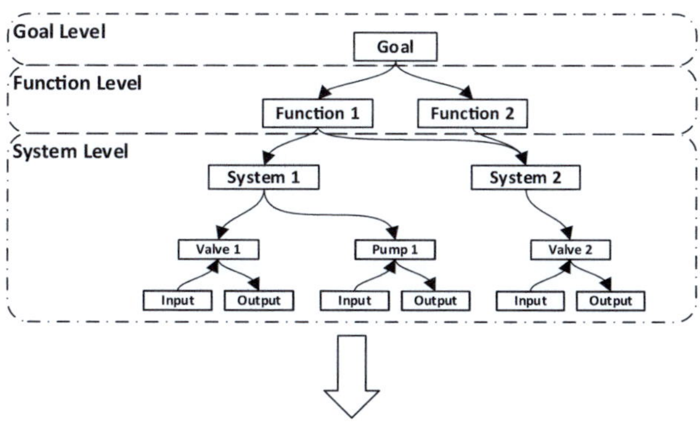

Goal	Function	System	Function redundancy	Component	Input Output
Goal	Function 1	System 1	Function 1, Function 2	Valve 1, Pump 1	System 1 input System 1 output
Goal	Function 1	System 2	x	Valve 2	System 2 input System 2 output
Goal	Function 2	System 1	Function 1, Function 2	Valve 1, Pump 1	System 1 input System 1 output

Fig. 2. Database form of function-based hierarchical framework

2.2 Multi-system Deep Learning Network

The multi-system deep learning network is applied to develop the algorithm for autonomous control. Deep learning network, which is a type of machine learning, requires careful engineering and considerable domain expertise to design a feature [11]. Deep learning networks are defined as the input/output values of the system analyzed in the function-based hierarchical framework.

A multi-system network integrates multiple deep learning networks modeling systems, based on the multi-modal deep learning [12]. It presents an integrated control structure for NPPs safety systems. The function and systems for the function organize a multi-system as based on the information analyzed in the function-based hierarchical

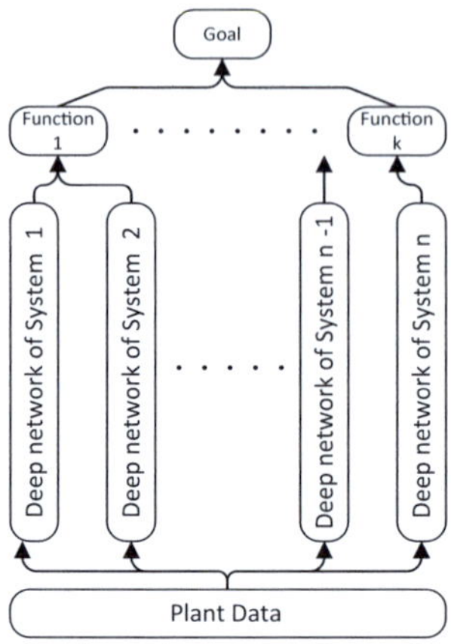

Fig. 3. Structure of multi-system deep learning network.

framework. Individual networks are trained using deep learning network in advance. Figure 3 is structure of multi-system deep learning network.

3 Modeling NPP Safety Systems for Multi-system Deep Learning Network

This section models NPP safety systems as a result of constructing the database of the function-based hierarchical framework and the structure of multi-system deep learning network. The structure of goal-function-system for the NPP safety is shown in Fig. 4.

The ultimate goal of hierarchical framework is the NPP safety. The NPP safety aims at preventing core damage and release of radiation the public [13]. In order to achieve the goal, nine safety functions are generally implemented in NPPs. Table 2 shows the nine safety functions and purpose of functions.

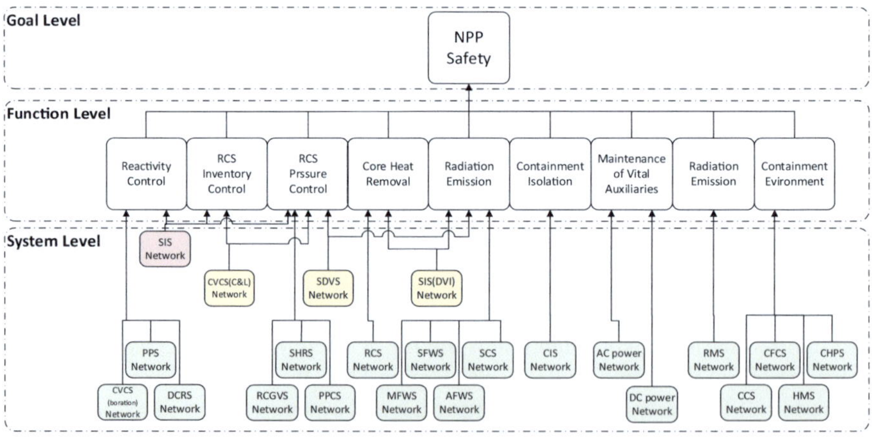

Fig. 4. Overall NPP safety system

3.1 Reactivity Control

The purpose of Reactivity control is to shut reactor down to reduce heat production. It consist of Plant Protection System (PPS), Digital Control Rods System (DCRS), Safety Injection System (SIS) and Chemical and Volume Control System (CVCS). Table 3 shows the database to define the modeling elements about Reactivity Control. The number of system redundancy indicates the number of safety function in Table 2.

Table 2. Safety functions and purpose of functions

Number	Safety function	Purpose of functions
(1)	Reactivity control	Shut reactor down to reduce heat production
(2)	Reactor coolant system inventory control	Maintain a coolant of reactor coolant system
(3)	Reactor coolant system pressure control	Maintain a coolant pressure of reactor coolant system
(4)	Containment environment	Keep from damaging containment
(5)	Containment isolation	Close opening in containment
(6)	Maintenance of vital auxiliaries	Maintain operability of systems needed to support safety systems
(7)	Core heat removal	Transfer heat from core to a coolant
(8)	Radiation emission	Control radioactivity to protect public
(9)	Reactor coolant system heat removal	Transfer heat out of coolant system medium

Table 3. Reactivity control

Goal	Function	System	System redundancy	System component	System input	System output
Prevent of core damage	(1) reactivity control	PPS	x	CEDM	Power, flow, temperature, pressure	Reactor trip signal
		DCRS	x	CEDM	Temperature, neutron flux, power	Rod control signal
		SIS	(1), (2), (3), (7)	SI pump, SI tank, SI valve	RCS pressure, RCS level, RCS temperature	Safety injection signal
		CVCS (boration)	x	Boric acid tank, makeup pump, makeup flow valve	Boron concentration	Make up flow valve control signal, Boric acid pump signal

3.2 Reactor Coolant System (RCS) Inventory Control

The purpose of RCS Inventory Control is to maintain a coolant of reactor coolant system. It consist of systems as Safety Injection System (SIS) and Chemical and

Table 4. Reactor coolant system inventory control

Goal	Function	System	System redundancy	System component	System input	System output
Prevent of core damage	(2) RCS inventory control	SIS	(1), (2), (3), (7)	SI pump, SI tank, SI valve	RCS pressure, RCS level, RCS temperature	Safety injection signal
		CVCS (charging and letdown)	(2), (3)	Charging flow control valves, letdown orifice isolation valves	RCS pressure, RCS level	Valve state signal

Volume Control System (CVCS) charging and letdown. Table 4 shows the database to define the modeling elements about RCS Inventory Control.

3.3 Reactor Coolant System (RCS) Pressure Control

The purpose of RCS Pressure Control is to maintain a coolant pressure of reactor coolant system. It consist of systems as Safety Injection System (SIS), Chemical and Volume Control System (CVCS) charging and letdown, Safety Depressurization and Vent System (SDVS), Reactor Coolant Gas Venting System (RCGVS), RCS Secondary Heat Removal System (SHRS) and Pressurizer Pressure Control System (PPCS). Table 5 shows the database to define the modeling elements about RCS Pressure Control.

Table 5. Reactor coolant system pressure control

Goal	Function	System	System redundancy	System component	System input	System output
Prevent of core damage	(3) RCS pressure control	SIS	(1), (2), (3), (7)	SI pump, SI tank, SI valve	RCS pressure, RCS level, RCS temperature	Safety injection signal
		CVCS (charging and letdown)	(2), (3)	Charging flow control valves, letdown orifice isolation valves	RCS pressure, RCS level	Valve state signal
		SDVS	(3), (9)	POSRV	RCS pressure	Valve state signal
		RCGVS	x	RCGV Valves	RCS pressure	Valve state signal
		SHRS	x	MSSV, ADVS	Steam generator pressure and level	Valve state signal
		PPCS	x	PZR spray isolation valve, PZR spray valves, Pressurizer heaters	PZR pressure and temperature	Valve state signal, heater signal

3.4 Containment Environment

The purpose of Containment Environment is to keep from damaging containment. It consist of systems as Containment Spray System (CCS), Containment Fan Cooling System (CFCS), Hydrogen Mitigation System (HMS) and Containment Hydrogen Purge System (CHPS). Table 6 shows the database to define the modeling elements about Containment Environment.

Table 6. Containment environment

Goal	Function	System	System redundancy	System component	System input	System output
Prevent of core damage	(4) containment environment	CCS	x	Containment spray	Containment pressure and temperature	Spray state signal
		CFCS	x	Fan cooler	Containment pressure and temperature	Fan state signal
		HMS	x	Hydrogen ignitors	Hydrogen concentration	Ignitors state signal
		CHPS	x	Hydrogen purge	Hydrogen concentration	Purge state signal

3.5 Containment Environment

The purpose of Containment Environment is to keep from damaging containment. It consist of systems as Containment Spray System (CCS), Containment Fan Cooling System (CFCS), Hydrogen Mitigation System (HMS) and Containment Hydrogen Purge System (CHPS). Table 6 shows the database to define the modeling elements about Containment Environment.

3.6 Containment Isolation

The purpose of Containment Isolation is to close opening in containment. It consist of systems as. Table 7 shows the database to define the modeling elements about Containment Isolation.

Table 7. Containment Isolation

Goal	Function	System	System redundancy	System component	System input	System output
Prevent of core damage	(5) containment isolation	CIS	x	Containment isolation valve	Leakage rate	Valve state signal

3.7 Maintenance of Vital Auxiliaries

The purpose of Maintenance of Vital Auxiliaries is to maintain operability of systems needed to support safety systems. It consist of systems as supply AP and DC power system. Table 8 shows the database to define the modeling elements about Maintenance of Vital Auxiliaries.

Table 8. Maintenance of vital auxiliaries

Goal	Function	System	System redundancy	System component	System input	System output
Prevent of core damage	(6) maintenance of vital auxiliaries	Supply AP power system	x	Emergency diesel generator (EDG), unit aux transformer, alternate AC diesel generator (AADG), S/by aux transformer	Load sequencing, voltage & frequency	Component state signal
		Supply DC power system	x	Station batteries	Voltage & frequency	Component state signal

3.8 Core Heat Removal

The purpose of Core Heat Removal is to transfer heat from core to a coolant. It consist of systems as Safety Injection System (SIS) DVI and Reactor Coolant System (RCS). Table 9 shows the database to define the modeling elements about Core Heat Removal.

Table 9. Core heat removal

Goal	Function	System	System redundancy	System component	System input	System output
Prevent of core damage	(7) core heat removal	SIS (DVI)	(7), (9)	Direct vessel injection (DVI) nozzle	Containment pressure	Nozzle state signal
		RCS	x	Reactor Coolant Pump (RCP)	RCS temperature and pressure	Pump state signal

3.9 Radiation Emission

The purpose of Radiation Emission is to transfer heat out of coolant system medium. It consist of systems as Release Path & Monitoring Control System (RMS). Table 10 shows the database to define the modeling elements about Radiation Emission.

Table 10. Radiation emission

Goal	Function	System	System redundancy	System component	System input	System output
Prevent of core damage	(8) radiation emission	RMS	x	Isolation valve, monitoring component	Radiation level	Valve state signal, monitoring data

3.10 Reactor Coolant System (RCS) Heat Removal

The purpose of RCS Heat Removal is to transfer heat out of coolant system medium. It consist of systems as Main Feed-Water System (MFWS), Start-up Feed-Water System (SFWS) Auxiliary Feed-Water System (AFWS), Safety Depressurization and Vent System (SDVS), Shutdown Cooling System (SCS) and Safety Injection System (SIS) DVI. Table 11 shows the database to define the modeling elements about RCS Heat Removal.

Table 11. RCS heat removal

Goal	Function	System	System redundancy	System component	System input	System output
Prevent of core damage	(9) RCS heat removal	MFWS	x	MFWP	RCS temperature	Pump state signal
		SFWS	x	SFWP	Manual signal	Pump state signal
		AFWS	x	AFWP	SG level	Pump state signal
		SDVS	(3), (9)	POSRV	RCS pressure	Valve state signal
		SCS	x	IRWST, POSRV	RCS pressure and temperature	Valve state signal
		SIS (DVI)	(7), (9)	Direct vessel injection (DVI) nozzle	Containment pressure	Nozzle state signal

4 Conclusion

This paper has discussed an approach to the autonomous algorithm to control the safety systems of NPP by using the deep learning network. NPP has a complex physical system and it is appropriate to use the hierarchical framework to analyze NPP safety system. This approach enables the systematic analysis of the power plant system and development of the database for the deep learning network. The developed database is to be applied to the deep learning network.

References

1. Nasri, S., et al.: Autonomous hybrid system and coordinated intelligent management approach in power system operation and control using hydrogen storage. Int. J. Hydrogen Energy **42**(15), 9511–9523 (2017)
2. Li, X., et al.: Development of a new integrated local trajectory planning and tracking control framework for autonomous ground vehicles. Mech. Syst. Sig. Process. **87**, 118–137 (2017)
3. Guo, H., et al.: Regional path moving horizon tracking controller design for autonomous ground vehicles. Sci. China Inf. Sci. **60**(1), 013201 (2017)

4. Petres, C., et al.: Path planning for autonomous underwater vehicles. IEEE Trans. Robot. **23** (2), 331–341 (2007)
5. Arora, A., Robert, F., Salah S.: An approach to autonomous science by modeling geological knowledge in a Bayesian framework. arXiv preprint arXiv:1703.03146 (2017)
6. Silver, D., et al.: Mastering the game of go with deep neural networks and tree search. Nature **529**(7587), 484–489 (2016)
7. Antsaklis, P.J., Passino, K.M.: Introduction to Intelligent Control Systems with High Degrees of Autonomy. Kluwer Academic Publishers, Berlin (1993)
8. Liu, Z., Fan, J.: Technology readiness assessment of small modular reactor (SMR) designs. Progress Nuclear Energy **70**, 20–28 (2014)
9. Ruan, D.: Intelligent systems in nuclear applications. Int. J. Intell. Syst. **13**(2–3), 115–125 (1998)
10. Sadighi, M., Setayeshi, S., Salehi, A.A.: PWR fuel management optimization using neural networks. Ann. Nuclear Energy **29**(1), 41–51 (2002)
11. LeCun, Y., Bengio, Y., Hinton, G.: Deep learning. Nature **521**(7553), 436–444 (2015)
12. Ngiam, J., et al.: Multimodal deep learning. In: Proceedings of the 28th International Conference on Machine Learning (ICML 2011) (2011)
13. Corcoran, W.R., et al.: The critical safety functions and plant operation. Nuclear Technol. **55** (3), 690–712 (1981)

Author Index

P. Fechtelkotter and M. Legatt (eds.), *Advances in Human Factors in Energy: Oil, Gas, Nuclear and Electric Power Industries*, Advances in Intelligent Systems and Computing 599, DOI 10.1007/978-3-319-60204-2

MIX
Papier aus verantwortungsvollen Quellen
Paper from responsible sources
FSC® C105338

If you have any concerns about our products,
you can contact us on
ProductSafety@springernature.com

In case Publisher is established outside the EU,
the EU authorized representative is:
Springer Nature Customer Service Center GmbH
Europaplatz 3, 69115 Heidelberg, Germany

Printed by Libri Plureos GmbH
in Hamburg, Germany